CICHLIDS
from Central America

by Ad Konings

Cichlasoma carpinte. This is a male. Photo by Hans Joachim Richter.

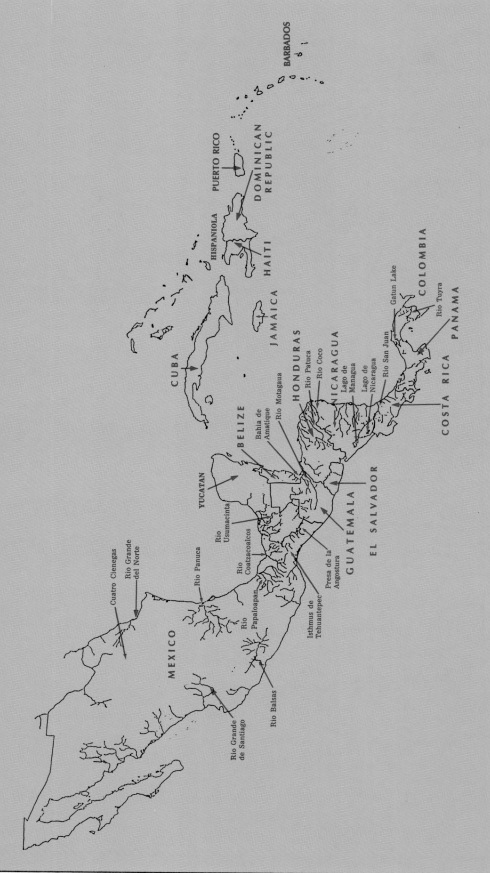

MAP OF CENTRAL AMERICA WITH SOME IMPORTANT RIVER SYSTEMS

C I C H L I D S
FROM CENTRAL AMERICA

Cichlasoma salvini. Photo by Werner & Stawikowski.

by Ad Konings

Distributed in the UNITED STATES by T.F.H. Publications, Inc., One T.F.H. Plaza, Neptune City, NJ 07753; in CANADA to the Pet Trade by H & L Pet Supplies Inc., 27 Kingston Crescent, Kitchener, Ontario N2B 2T6; Rolf C. Hagen Ltd., 3225 Sartelon Street, Montreal 382 Quebec; in CANADA to the Book Trade by Macmillan of Canada (A Division of Canada Publishing Corporation), 164 Commander Boulevard, Agincourt, Ontario M1S 3C7; in ENGLAND by T.F.H. Publications Limited, Cliveden House/Priors Way/Bray, Maidenhead, Berkshire SL6 2HP, England; in AUSTRALIA AND THE SOUTH PACIFIC by T.F.H. (Australia) Pty. Ltd., Box 149, Brookvale 2100 N.S.W., Australia; in NEW ZEALAND by Ross Haines & Son, Ltd., 18 Monmouth Street, Grey Lynn, Auckland 2, New Zealand; in SINGAPORE AND MALAYSIA by MPH Distributors (S) Pte., Ltd., 601 Sims Drive, #03/07/21, Singapore 1438; in the PHILIPPINES by Bio-Research, 5 Lippay Street, San Lorenzo Village, Makati Rizal; in SOUTH AFRICA by Multipet Pty. Ltd., 30 Turners Avenue, Durban 4001. Published by T.F.H. Publications, Inc. Manufactured in the United States of America by T.F.H. Publications, Inc.

TABLE OF CONTENTS

Map of Central America
Cichlids in Central America, 7
Central American Cichlids in the Aquarium, 11
   Cichlids need their own aquarium, 11
   The aquarium, 13
   Decoration,14
   The water, 17
   Biological filtration, 20
   Stocking the aquarium, 25
   Food, 28
Feeding Specializations in Central American Cichlids, 31
   Primitive predators from Caribbean islands, 31
   Primitive piscivores from the mainland, 33
   Snail-crushers and leaf choppers, 41
   Primitive vegetarians, 44
   Specialized herbivores, 47
   Invertebrate-feeders from the subgenus *Theraps*, 56
   Small predators from the subgenus *Archocentrus,* 59
   Large invertebrate-pickers, 62
   Invertebrate pickers from the South, 66
   Substrate-sifting insectivores, 67
   Substrate-sifting species from the North, 69
   Substrate-filterers from shallow water, 69
   Cichlasoma from South America, 73
Breeding Central American Cichlids, 77
   Pair formation, 77
   Finding a place to breed, 78
   Concealing eggs, 80
   Eggs and hatching, 80
   Feeding fry, 85
   Defense of the fry, 86
Representative Central American Cichlids, 89
   Cichlasoma (Theraps) synspilus, 89
   Cichlasoma (Theraps) maculicauda, 99
   Cichlasoma (Theraps) intermedium, 107
   Cichlasoma (Theraps) hartwegi, 109
   Cichlasoma (Theraps) fenestratum, 111
   Cichlasoma (Theraps) lentiginosum, 113
   Cichlasoma (Theraps) nicaraguense, 114
   Cichlasoma (Theraps) sieboldii, 118
   Cichlasoma (Tomocichla) tuba, 119
   Neetroplus nematopus, 121
   Cichlasoma (Archocentrus) nigrofasciatum, 123
   Cichlasoma (Archocentrus) centrarchus, 130
   Herotilapia multispinosa, 134
   Cichlasoma (Amphilophus) longimanus, 142
   Cichlasoma (Amphilophus) alfari, 146
   Cichlasoma (Amphilophus) citrinellum, 150
   Cichlasoma (Amphilophus) altifrons, 158
   Cichlasoma (Thorichthys) aureum, 160
   Cichlasoma (Herichthys) carpinte, 164

The Ancient Continents, 168
Evolution of Central American Cichlids, 169
Illustrated Atlas, 176
Checklist of Central American Cichlids, 218
Bibliography, 221
Index

This is a photograph of the type specimen of the fossil *Cichlasoma woodringi* Cockerell, 1923. It was loaned to the photographer, Dr. Herbert R. Axelrod, by the Smithsonian Institution, Washington, D.C.

When Cockerell described a fossil cichlid from Haiti in 1923, it became evident that cichlids were present in Central America at least 25 million years ago. At that time Haiti was not part of an island in the Caribbean Sea but was most likely located where we find Nicaragua today. Then it was part of an island bridge between South America and what is called Nuclear Central America. These islands facilitated the migration of South American and, perhaps, African cichlids toward the northern section of Central America.

Cichlids belong to a very advanced and successful fish family. They are sometimes spoken of as intelligent fishes, but it is not so much their "intelligence" but more their anatomical features that make

the cichlids very successful colonizers of new areas. It has been demonstrated on several occasions that cichlids are among the first to occupy newly formed or available habitats. The famous Great Lakes of Africa, for instance, have a fish fauna dominated by cichlids. They possess a great capacity to adapt themselves to a new or a changed environment. Like the water in most African lakes, the fresh waters in Central America contain a considerable amount of minerals, favoring the presence of cichlids. Cichlids are so-called secondary freshwater fishes, which means that they originated from marine species. The marine ancestry is still expressed in the usual cichlid tolerance of high concentrations of minerals, and some species may thrive in pure

sea water. Even today some Central American cichlids are capable of living in the Caribbean Sea: *Cichlasoma maculicauda*, the most widely dispersed cichlid from Central America, may be encountered among the coral reefs of the seashores. Through geographic isolation and other evolutionary processes, the first few cichlid species that arrived in Nuclear Central America developed into the considerable number of species that are recognized today.

Cichlids adapt readily to artificial environments such as tap water and glass tanks, and the popularity of cichlids increases among aquarists every year. This fact is born not only from the ease with which these fishes can be maintained, but also from their behavior, which is readily displayed in captivity.

CICHLIDS IN CENTRAL AMERICA

Central American cichlids are very colorful and hardy fishes that are substrate breeders that defend their offspring in a most demonstrative manner. In the usual course of affairs, a pair select a spawning site that might be in the open water or hidden in a cave and the site is cleaned of any loose material. Before the actual spawning takes place, some ritualized behavior is displayed. As is typical for American cichlids, the fish may grasp each other's mouth and start some tugging and jerking. During this prelude jawbones may get broken or even ripped off, but the pair still may "love" each other and decide to spawn shortly afterward. When the eggs are deposited the pair move in circles one behind the other. The female deposits a row of up to 20 eggs that are then fertilized by the male, who travels immediately behind his spouse.

After two to three days the eggs hatch, the fry being chewed free from the eggshells. After another three to five days the fry are mobile. This is a most enjoyable period, watching a large cloud of fry surround the

Mouth tugging is a typical characteristic of many American (South American) cichlids. A labyrinth fish from Asia, *Helostoma temmincki*, the kissing gourami, also indulges in "kissing." No one has offered an acceptable explanation for mouth tugging. If you look closely there is always one fish that is biting the other. They are not both biting at the same time. Photo below by Uwe Werner. On the facing page are a duo of hybrid male *Cichlasoma*, also mouth tugging. Photo by Ruda Zukal.

guarding parents. For the first few weeks the young cichlids are dependent on planktonic food, which can be replaced satisfactorily by newly hatched brine shrimps. Soon the youngsters venture farther away from their parents, who see their task outgrowing their ability.

In the wild, juveniles may be protected by their parents for more than three months, then they are abandoned. At that moment the juveniles suffer the fiercest predation. In captivity, however, most of the fry will be eaten in the first weeks if left with the parents in the community tank. This is due to the overcrowded situation that is prevalent in most, if not all, aquaria. As a result of this rapid loss of offspring, the pair may breed again shortly afterward, usually within three months.

Most Central American cichlids acquire a breeding dress, nuptial coloration. This color pattern is different from the regular pattern and is an indication that spawning is to be expected soon.

There are spawns . . . and there are spawns. This pair of *Cichlasoma managuense* raised more than 1,000 babies in this spawn, which took place at the University aquarium in Nancy, France. Photo by Dr. Denis Terver.

CICHLIDS IN CENTRAL AMERICA

More examples of mouth tugging. Look closely and you'll notice that the fish really bite each other. In the photo above a *Cichlasoma managuense* is locking jaws with an *Astronotus ocellatus*. Photo by David Sands. Below: A pair of *Cichlasoma synspilus* are mouth tugging. Photo by Rainer Stawikowski.

The breeding colors are usually darker, with a great part being black markings, and they are not necessarily more enticing than the regular attire. On the one hand, the darker colors may be a signal for the fry, but on the other hand, they may warn neighboring cichlids to stay away.

There is not a single cichlid that is not worth observing or breeding. American cichlids are, besides their interesting behavior, very decorative. Moreover, it is not too difficult to keep a few different species together in a community aquarium. We only have to observe some rules that are not at all difficult to keep.

There are several ways to keep fishes. Personally, I prefer a natural tank in which the fish plays the most important part of the decoration. The bright colors of most American cichlids stand out nicely against a dark, inconspicuous background. A colorful background with various types of decorative material will never be in harmony with the inhabitants, since in such tanks attention is diverted from the fish. Since the conspicuous background is static, it might become boring in a short while. In a "natural" tank you never get bored watching cichlids. Even friends and family who are not interested in fishes at first are constantly attracted to the ever-changing movements of cichlids.

Cichlids need their own aquarium

Many cichlids from Central America grow to a respectable size and as such are in need of ample space. An important point is that the tank must be decorated and set up especially for cichlids, even when other fishes are added. The substrate spawning behavior of these cichlids calls for territories that should be staked out easily. Therefore, you should know how many and what species you are going to keep in your tank. Typical Central American species will be discussed later to help you make a choice. For now it is important to realize that a pair of 30 cm (1 ft) *Cichlasoma* need a territory that has a diameter of at least three times their length. Under natural circumstances the territory is of course much larger. When several equally large pairs are kept together, territorial areas decrease in relation to the fish's length. This is a general rule that cannot be applied blindly to all species. The most important fact that we have to tackle is the

Cichlasoma atromaculatum, as well as other Central American cichlids, requires a large aquarium with lots of stones and caves. Photo by A. Roth.

Cichlasoma octofasciatum. **Photo by Hans Mayland.**

aggression of cichlids. It is entirely possible to maintain a harmonious community consisting of several species for a long period.

The Aquarium

After you decide which species you want to keep and what size tank you should have, you have to find the right place to put it. It is advisable to place an aquarium in the living room or some other spot where you can relax in front of it. Children are as a rule very attracted to fishes, so do not place it too high. The larger the tank the lower it should be placed. A 1000-liter (250 gal) tank seems less massive when 30 cm (1 ft) from the floor than when on a 90-cm (3-ft) table. Moreover, it will neither dominate nor shrink the interior of the room. The aquarium should not receive direct sunlight, at least not at midday, or the algal growth on the glass might become considerable and call for a daily clean-up.

About the size of your tank: don't make it too small! The minimum size is one meter (3.3 ft), but a larger size is recommended. It sounds contradictory, but the bigger the tank the easier is its maintenance.

To not only support the contemporary topic of saving energy, but also to relieve your monthly electricity bill, you should insulate the tank. A wrapping of 2.5 cm (1 in) styrofoam will reduce the heating costs of your tank by a factor of two when placed in a heated room; in a cellar or other barely heated environment it may save you much more. Today all-glass aquaria are the most popular. Place the aquarium on a layer of styrofoam situated between the bottom glass and the table or other support. To comfort the fishes, the back and at least one

There is no tank designed exclusively for Central American cichlids, but the principles are firmly established depending upon the results you wish to achieve. If you want to keep many large cichlids, then of course you need a very large tank. If, on the other hand, you only want to keep one pair and you want to spawn them and raise the young with the parents, then a much smaller tank will suffice. When you buy the tank for a solitary pair you must consider how large the fish will grow. A tank about 40 inches (one meter) in width is ideal.

side of the tank should be covered. If only the front is exposed, we still can watch, yet the inhabitants will feel more at ease since more hiding places are available. As a result, the fishes will be more visible and will spawn more easily. Normally the aquarium is equipped with a hood in which some incandescent or fluorescent bulbs are installed. The light is not of great important for the cichlids, but it is needed for plants, if present.

Decoration

The most important part of the decoration is the background. The simplest solution is to paint the outside of the back glass. Pick dark colors, anything ranging from deep red to brown, green, deep blue, or black. Do not try to create a van Gogh when mixing colors. In the long run a monochromatic, dark background satisfies best. If available, you can fix a plate of non-toxic, inert, colored plastic to the inside back glass. More frequently available, but not advisable, are colored styrofoam plates. Cichlids nibble and bite anything, including your newly installed styrofoam background, and the result is white-speckled scenery that draws your attention and annoys you.

As an alternative you could place the plastic plate outside the tank. The drawback, like that of painting the back glass, is that it will yield reflections from behind that reduce the natural appearance of your tank. A

background made of sheets of compressed cork is a reasonable but not long-lasting solution to keep the wallpaper from being visible through your aquarium. A "weightier" solution for the background could be stone slabs. Stones and also other artifacts can be fixed to glass by the same silicone cement used to glue the tank glass. Although slates or stone may give a natural look, they are difficult to match to form a completely closed background. An additional painting of the outside (with the same color as the stones) might help you out.

Stones, rocks, and slabs are widely used decorative materials that usually are not toxic and are always available. We make a mistake if we just dump a heap of stones in the tank. Although there will be ample necessary hiding places, there usually is no pleasant formation to look at. To make the tank seem bigger you should use large rocks and not fist-size stones. The heavy weight of such rocks prevents their use

Above and below: This schematic shows a setup suitable for cave dwelling dwarf cichlids. **H** = *Hygrophila*. **S** = *Sagittaria*. **I** = *Ceratopteris*. **AU** = *Aponogeton undulatum*.

Above and below: This schematic shows another setup suitable for a cave dwelling dwarf cichlid. **C** = *Cabomba*. **V** = *Vallisneria*. **CB** = *Cryptocoryne becketti*.

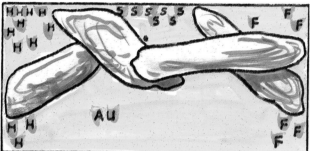

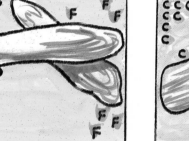

Above and below: This schematic shows a suitable layout for an open spawning dwarf cichlid. **L** = *Ludwigia*. **SP** = *Echinodorus intermedius*. **V** = *Vallisneria*.

Above and below: These two drawings are further layouts for a suitable open spawner tank. **V** = *Vallisneria*. **CC** = *Cryptocoryne cordata*. **M** = *Myriophyllum*. **HG** = *Eleocharis*.

Depending upon the type of spawner (cave, open, etc.), you will want to decorate the tank with a few very hardy plants growing from a container, some rocks and a large piece of driftwood. Buy the driftwood at a petshop, not at a garden store, as petshops are careful about the source and treatment of the driftwood.

Cichlasoma atromaculatum male. Below: This tank is ideal for most Central American cichlids. It has a perfect background, and the rocks allow surfaces for open-spawners.

on glass bottoms. In my show tank a rock of approximately 250 kilos (550 lbs) dominates the decoration. Many visitors wonder how the glass could support this large stone . . . until I tell them that it is not a real stone but a ceramic one! The great advantages of ceramic stones are their solidity and lack of weight. The artist who constructed these stones had to punch holes in the top otherwise they would float to the surface! With patience, you could make your own stones in the shape you want and have a friendly potter bake your rocky-coast imitation. Stones and other heavy materials should be placed on the bottom before sand or gravel is added to the tank. This prevents a collapse when cichlids dig the sand out from beneath the rocks. Use a thin sheet of styrofoam to protect the glass bottom. Large flowerpots can be hidden by stones and serve as caves into which the cichlids may retreat.

After you have fixed the background and the heavy decorations, the bottom is covered with sand or gravel. Do not use any other material, such as peat moss or garden soil. Washed fine sand looks very natural but is not advisable when you want to put plants in the aquarium. You are better off using some fine granulate gravel. Sand has the disadvantage that dirt remains on top, while the structure of gravel allows the dirt to settle between the particles.

After the tank is filled with water, the other decorations are given their places in the underwater landscape. Driftwood can be nicely combined with stones. Plants are difficult to use with Central American cichlids because the fishes like to redecorate their homes themselves. They won't move rocks, but plants usually are victims of their unmistakable taste for redecorating an aquarium. Heaters and other fragile equipment should be avoided in the tank itself or affixed in such a manner that big *Cichlasoma* will not smash them during one of their housecleaning bouts.

Besides plastic plants, a genuine waterplant, *Ceratophyllum demersum*, could be used with these cichlids. This waterplant has no roots and can be bundled together. Weighted with a pebble, the plant can be put between rocks conveniently. A plant without roots can never be uprooted.

The water

It must be clear that a natural tank does not exist. The usual nature of the home-waters of many cichlids is not to our liking for a tank in the living room. Crystal-clear water rarely is found in Central America, but visibility in

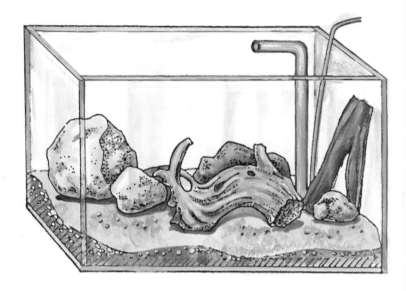

This type of tank is a commercial breeder's setup, inexpensive and easy to care for and clean. An undergravel filter, small pebbles and sand, rocks and suitable branches of driftwood (which you must only buy from your aquarium store)   form small caves. Below: This is a more elaborate setup. A large flat stone is for open-spawners. The large rock can be used as a cave when the breeders dig one, or as a surface for open-spawners. The living plants give it a suitable appearance for the living room, and the terraced undergravel filter keeps the gravel from slipping too much.

This is an ideal aquarium setup for such fish as *Herotilapia multispinosa* that like to spawn in the open on a flat rock, then hide their young among dense plant roots and leaves.

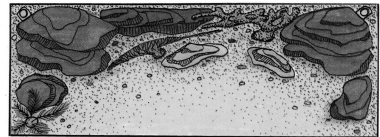

some streams may be more than 2 meters (6.6 ft). Murky green or dark water is more prevalent and may harbor the most colorful species. Many American cichlids live in well-oxygenated running water that is almost devoid of any organic waste or toxic products. We have to bear in mind this fact when we install a water circulating and filtering device. It is not the air bubbles but more the movement of the water that brings the necessary oxygen. A

A top view shows how rocks can be arranged for beauty as well as utility for the Central American cichlid tank. Note that the rocks form a three-sided background. This keeps shadows and other disturbances out of the tank. Left: If you get a large enough piece of driftwood, you will be able to cut holes in it and thus provide caves for the cave-spawners.

clearly visible air releaser in the tank is unnatural and will churn up the dirt and deposit it on the decorations.

Large cichlids produce a lot of waste that we have to remove efficiently. A power filter circulates the water and mechanically removes the waste. You have to know the pH (a measure of the relative acidity or alkalinity) of your tap water. In any aquarium shop you can buy an indicator that will quickly tell you the pH of the water. Cichlids from Central America have a very wide tolerance regarding the pH of water, and something between 6.5 and 8.5 will do fine. The amount of minerals in the water, measured as conductivity and not as hardness, is irrelevant as long as you are not using water from the Great Salt Lake. A low pH, below 6.5, might cause some breathing difficulties in American cichlids; try to avoid this condition.

Above right: There is a great deal of skill involved in feeding cichlids—or any other aquarium fish, for that matter. Just dumping food into a tank is, of course, unsatisfactory. You must feed very small amounts until you have an accurate measure of the requirements of the fishes. Keep in mind that it is much safer to err on the side of providing too little food than on the side of feeding too much. Many experienced aquarists even dwarf their fishes purposely. Then they can fit more into a tank. Dwarfed fish have normal-size spawns. Above left: Small fishes such as some of the smaller livebearers can be maintained in a small tank with a bottom filter and used as food for the larger cichlids. If you feed the smaller fishes heavily before you feed them to the cichlids, the cichlids will benefit.

There is a large philosophical discussion among cichlid breeders about the advantages of pellet foods over flake foods. Pellet foods are best because they float longer and therefore you can better judge how much food is uneaten. Floating pellets that are uneaten can easily be removed with a net. The best pellets sink very slowly and are manufactured so some pellets float much longer than other pellets. This is usually done by the addition of fat during the manufacturing process.

Biological filtration

In a large aquarium (over 400 liters or 100 gallons) it is advisable to install a biological filter. This filter can be combined with a mechanical power filter or be separately operated. Body wastes, decayed plant leaves, and deceased fish burden the water with organic metabolic products. The most toxic substance among these products is ammonia. As an ammonium salt it is relatively harmless, but in a gaseous state ammonia is very toxic. Fortunately the gas dissolves poorly in regular aquarium water, but if the pH reaches around 9 and ammonia is present, it will lead to a disastrous death of all tank inhabitants. In a well-maintained aquarium in which regular water changes are made, such problems are not encountered.

The ammonium salt, however, could give some problems and you have to tackle these.

Under natural conditions these ammonium salts are processed by denitrifying bacteria. Chemically changed to nitrite and nitrate they become a harmless substance that will be incorporated into plants and algae. In overcrowded tanks, freshly setup tanks, and in tanks where the complete bacterial population has been wiped out by a recent antibiotic treatment, the ammonia will pile up. The growth of a new population of denitrifying bacteria could take several weeks or months. These very important bacteria need oxygen, which usually is not abundant in overcrowded tanks. This may further hamper the stabilization of a healthy environment for your cichlids. In new tanks (or tanks

deprived of bacteria) the rising concentration of ammonia could trigger an explosive growth of blue-green algae. These primitive plants can use the ammonia directly and successfully grow in these unstable conditions. An explosive growth of blue-green algae is a sure sign that the nitrogen cycle has not stabilized yet. Do not use any chemicals to remove the velvet green layers—it is better to leave them in the tank as they at least will absorb the ammonia. Frequent water changes (50% twice a week), small portions of food, and a

---

It is important for all aquarists to understand the nitrogen cycle. Read this section carefully and use the diagram below to help you better understand it. Mysterious fish deaths are often attributed to water poisoning.

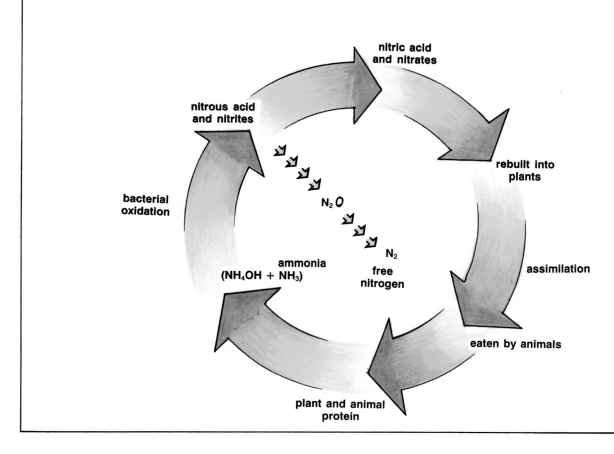

certain amount of patience are all that are needed to rid you of the plague. A few bunches of real waterplants, like *Vallisneria* and *Ceratophyllum*, might speed up the process. Until a healthy population of denitrifying bacteria is present, the blue-green algae will thrive. At a certain moment the velvet green algal layers get a little darker and might show some brown spots. At this time you could mechanically remove it.

Always try to avoid overfeeding. The untouched food will collect at certain spots and suffocate the bacterial fauna underneath and gradually in the whole tank.

The denitrifying bacteria need a substrate and lots of oxygen to operate. Usually they are located in the gravel and the sand. It is not recommended that you use an undergravel filter. The amount of waste large cichlids produce combined with the digging

activities employed by many of them rule out the effective usage of such a filtering system. Moreover, if you ever treat your cichlids with antibiotics, the denitrifying bacteria in the undergravel filter will be killed. Of course, it would be better to treat a diseased fish in a separate tank but if all inhabitants have to be treated, the quarantine tank would be overcrowded and certainly not enhance the treatment. For the price you have to pay for a sophisticated undergravel filter you could buy a small tank and start a biological filter yourself. This filter is connected with the main tank. The biological filter contains ample substrate to support a much bigger population of bacteria. Moreover, during an antibiotic treatment it can be disconnected from the main tank and be operated in closed circuit. Further advantages are that all

This is a commercial type of biological filtration. The water from the aquarium enters the first of four chambers at point **A**. This is a settling tank in which large particles fall to the bottom of the first chamber. The water overflows the first chamber at point **B** and enters the first of two biological filters in which bacteria break down harmful aquarium by-products. The top of this second chamber has a filter material to remove the larger particles that might hamper water flow. This must be changed frequently. This chamber is filled with almost any non-toxic substances that have a large surface area. Lava stone, hair-curlers, Bio-balls, and small gravel are all satisfactory. The water drops by gravity from the second chamber to a reservoir at **C**. This is a second settling chamber. The water then flows from the settling chamber **D** into the second biological filter chamber **E**, which is also filled with lava stone, etc., and finally is pumped back into the aquarium at point **F**.

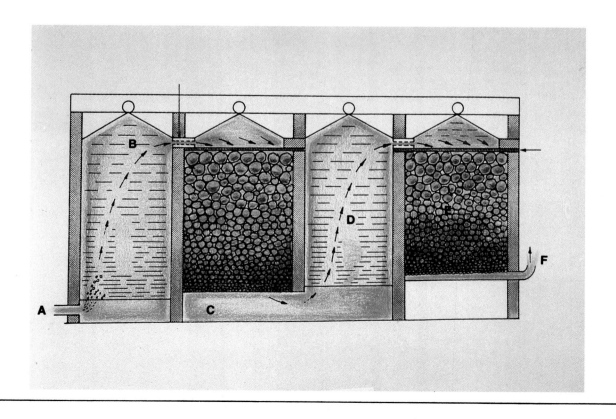

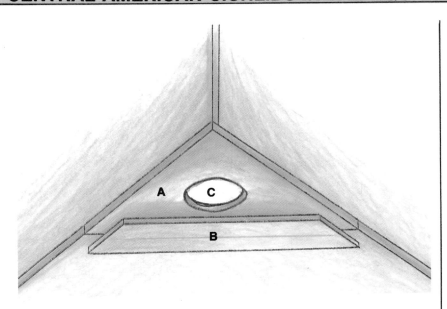

A very efficient biological filtration system involves cutting a small corner from the bottom of the tank as shown in **A** above. If you are able to drill a hole in the bottom of the tank, like **C**, then cutting off the corner is not necessary. The corner removed is replaced with a piece of plexiglass of the same dimensions, and this piece of plexiglass has a 5 cm (2 inch) hole drilled in it with a normal wood drill. The plexiglass is glued in with silicone adhesive that petshops sell for building aquariums. Another piece of glass is glued with silicone over the joint to make the joint more stable. This support is labeled **B**. Through this hole you make a connection with a biological filter (see previous and facing pages). Below: A side view of above in which a dam is constructed so the water falls over it. Filtering material can be placed in this chamber marked **D**.

Automatic water changers are invaluable for aquarists involved in adding, removing or filling water in their tanks. Make sure you get a plastic water line as long as necessary so it reaches comfortably from your tank to the faucet. It is also useful, by the way, for watering houseplants.

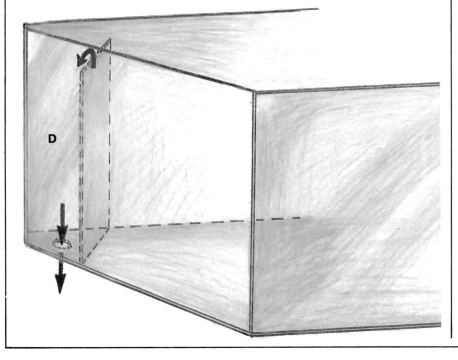

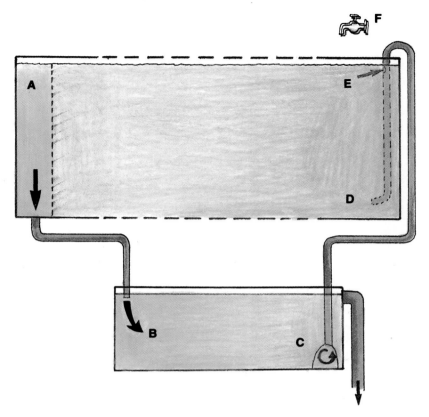

This is a continuation of a theoretical biological filter from the facing page. Water from the display tank spills over the dam **A**. It is piped to the bio-filter **B**. It is then cleaned and returned by a filtering power pump **C**, which pumps it back into the showtank **D**. A small hole, **E**, in the outlet prevents backflow during a power failure or pump failure. Changing water is made simple using the water changer (facing page) together with a faucet **F**.

AN OPEN BIOLOGICAL FILTER.
For illustrative purposes, the front pane of glass has been removed (plexiglass can also be used). The water from the display tank collects in **A**. From there it flows through the three substrate compartments (**B1, B2, B3**) where the denitrifying bacteria are established. The flat overflows (**C1, C2, C3**) allow for maximum oxygenation of the water. The semicircular shape facilitates access to the substrate upon which the bacteria are growing. From compartment **D** the cleaned water is pumped back to the display tank.

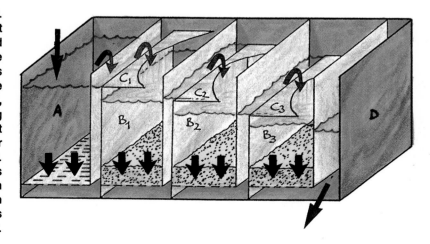

A powerhead attached to an undergravel filter aerates as well as increases the filtering capacity of the filter.

equipment, like a heater, airstones, and a mechanical filter, can be adequately incorporated into the filter. A disavantage could be that it needs a place to set. Usually it is put under the main tank, but if you reserved that space for something else (hopefully not your expensive hi-fi installation), you could put it beside or behind the main tank even in another room. The water from the main tank has to flow, by gravity, to the biological filter. From the filter it is pumped into the main tank again.

There are many biological filtering systems available at your pet shop, and I have tried several of them. The most efficient and silent system involves the following steps. A small corner is cut from the bottom. This is not necessary when you are able to drill a 5-cm (2-in) hole in the glass. The cut-out corner is replaced by a piece of plexiglass with similar dimensions. In this piece of plexiglass a 5-cm hole is drilled with a conventional wood-drill. The plexiglass is glued in place with silicone cement. To stabilize the corner a small strip of glass is glued over the joint. Anyway, there will hardly be pressure on it during operation. A short plastic pipe is glued (with silicone cement) into the 5-cm hole. This will be the connection with the filter. Inside the tank a plate of glass of the same thickness as the sides is glued obliquely in the corner. Be sure that the fitting is watertight. The height of the glass plate will determine the water level in the tank. When water is pumped into the aquarium it will flow out automatically when the level of the glass plate has been reached. The space behind the dam is filled with filter wool. The wool mechanically removes the visible dirt and should be rinsed or replaced when dirty, depending on the number and size of the fish. Plastic pipes will carry the pre-cleaned water to the

biological filter. Such a filter can be constructed by you or bought from your dealer. It is rather simple and does not have to look professional, since it is hidden under the tank.

For optimal effectiveness the filter chambers can be combined and used in series. A large number of fish in a tank calls for a large biological filter. For up to six 20-cm *Cichlasoma* two substrate chambers will suffice. Ten mature cichlids of a similar size need a three-chamber filter. The most important feature of the filter is that the water comes into close contact with the air before it enters the next chamber.

In contrast to older style filters, where the chambers had underwater connections, this filter optimizes the oxygenation of the water, which is very important for its processing by the bacteria. The first chamber can be used as an additional mechanical filter. It will ensure an unclogged filter substrate for at least a year. The substrate consists of a porous material to greatly extend the surface upon which the bacteria will settle. Several types can be purchased from the pet shop. Do not use sand or small gravel. The layer of substrate should not be higher than 6 to 8 cm (2½ - 3 in), as the oxygen will not penetrate greater thicknesses and additional substrate would not support bacteria in quantities.

In the last compartment an airstone and a combination heater could be installed. The cleaned water is best pumped out with an underwater pump, available in most pet shops. A similarly setup aquarium has only the pump outlet to be hidden away; all other equipment is safely put in the filter.

Since the inflowing water is well-oxygenated, it does not have to "rain" into the tank. An outlet close to the bottom of the tank, at the opposite side of the overflow, will effectively swirl all dirt from the gravel. A biological filter does not have to be cleaned until the filter is clogged with dirt. In fact, it could operate for years without servicing. You just have to take care that the filter wool is changed in time.

One thing you have to bear in mind: when the pump outlet is underwater a power failure may reverse the water current and flood your filter and perhaps your room. This problem is solved by constructing an overflow in your filter and connecting it directly with a sink or similar drain. A small hole drilled just below the surface in the outlet will draw air when water flow is reversed. This prevents the complete draining of the tank. When you have built a connection to the drain, water changing is really easy. My show tank is connected directly with the warm and cold mains. Changing water simply means turning a faucet on and closing it 15 minutes later, when 350 liters of water have been replaced. Excess water automatically runs into the drain.

Sponge filters are less sophisticated and also less effective. These filters will satisfy most needs in biological filtering, but their drawback lies in their size and the difficulties of hiding them behind decorations. For breeding tanks and other non-show tanks they are the best filters. They are quickly rinsed and easily installed.

Stocking the aquarium

When you have set up a tank as described above, you don't need to accompany the cichlids with open-water fishes. Since ample shelter is available, your cichlids feel safe and frequently venture out of their caves. Only in tanks where all sides are exposed and decorations are

Top: Various excellent air pumps are available from your local aquarium store. These air pumps may attach to an undergravel filter (center photo), an outside filter or a bottom filter. These small, inexpensive air pumps may be used for many other aquatic purposes as well as for constantly stirring food preparations, paints, etc. A more expensive, though more powerful, solution to the problem is the motor driven outside filter. These filters filter up to 25 times more water than an undergravel filter with an air pump. They are usually more quiet as well.

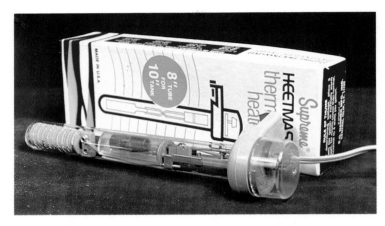

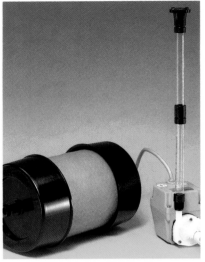

Sponge filters, like this Supreme Poolmaster by Danner, can be used for emergencies or as a constant filter for the cichlid tank. It is ideal for free-swimming fry that might otherwise be sucked into other types of filters. The blue sponges are easily cleaned or changed.

Top: A suitable heater-thermostat is necessary in most temperate zones to maintain a satisfactory water temperature for Central American cichlids. Middle: Power heads are available for attachment to undergravel filters to move much greater quantities of water, increasing the efficiency of the filters and giving movement to the water. Below: Many types of outside, hanging filters are also available at most petshops, including the outside power filter. A good petshop should have many different varieties of filters to show you.

very sparse may cichlids hide in any little corner they find. Central American cichlids are territorial substrate-breeders. As such they need a lot of space. Once they have staked out their territories it is difficult to add more fish to the tank. Therefore all inhabitants are placed into the tank at one time. Since all of them are entering a new environment, there is no fish claiming a certain territory. In the following days a kind of hierarchical order is set up among the inhabitants. Every fish or pair marks its territory.

If after one month no harmony exists in the group, some changes have to be made. Probably there are too many fishes or the species chosen resemble each other too closely. A cichlid community must be carefully chosen. The species must not look alike, as this will evoke aggressive behavior. Usually cichlids are aggressive against conspecifics since they mean competition, but visually

Most petshops sell a complete aquarium startup kit with tank, heater, filter, pump, etc., included in a special price structure.

and behaviorally similar species would also evoke a response. A wider variety in external morphology results in a better harmony among the inhabitants.

Under normal circumstances the newly introduced cichlids will not spawn the next day and therefore may not be in immediate need of a territory. When the time has come for them to spawn, though, it may present some problems. In the vocabulary of Central American cichlids, "territory" means "a place free of anything having fins." In an overpopulated aquarium it could trigger a disastrous fight among several inhabitants. If you have the feeling that your tank is not overpopulated, i.e., the water retains its crystal-clear appearance, but there are still fights, you could resort to an old trick: lower the temperature. The

Water tests kits are a must for any serious hobbyist.

Most filter manufacturers supply disposable cartridges to take the work out of cleaning a filter. These filter bags are usually washable, too.

You can attach a special water changing device to your faucet and change, add or fill water with it. Once you get accustomed to using these devices, you'll never have to carry buckets of water again. There are several kinds of water changers on the market.

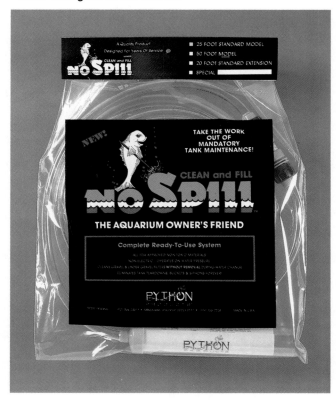

usual temperature should be around 25°C (77°F), but during periods of aggression it could be dropped to 20°C (68°F) without any harm being done. After you have literally cooled down the fish for two weeks, the temperature could be raised again. If similar fights occur you will have to remove the culprit. You might exchange it for a smaller individual or another species.

It takes some time before Central American cichlids develop their full coloration. Don't be impatient and don't change fish every month. Every introduction causes trouble, and an unsatisfactory situation may be created. Give these cichlids some time; under normal aquarium conditions they will develop the expected colors.

Food

There is nothing easier than feeding Central American cichlids. Their voracious

Petshops carry large varieties of food for Central American cichlids. Most Central American cichlids eat anything . . . or perhaps "everything" might be a better word. Just look at the large mouth which can be extended to suck in small fish (below).

There are many foods available. You should use as many different types as possible, thereby providing a good variety of nutritional substances. Keep in mind that the biggest danger involved in feeding your fishes is overfeeding. OVERFEEDING IS DEADLY. When you feed too much at one time the excess food just lies on the bottom of the tank and rots, causing the pollution of your tank and, eventually, a bad smell. Even those dry foods that won't cloud the water if they are fed in moderation definitely will cloud the water if you overfeed. On the right is a photo of *Gammarus,* a small crustacean that is among the best of all live foods for cichlids.

appetites settle for anything edible thrown into the tank. Many species are herbivorous, but others are insectivores or piscivores. In a community tank several types normally are present, and they have to be fed accordingly. Fortunately the pet industry provides us with many types of dried foods that will satisfy the needs of herbivores and invertebrate-feeders. Large piscivores (fish-eating predators) prove to be the real gluttons of the cichlid family. The predatory cichlids from Central America are in this respect no exception, but they are among the easiest to satisfy with dry pelleted food.

If you want to breed these guapotes (a common name for the large *Cichlasoma*) you will have to do more than the twice daily shake of the dried food container. They need some juicy meat. This does not necessarily mean that it has a better quality, but it just tastes better, so they will eat more of it. The extra ingested food indirectly stimulates the fish to breed. Beefheart is a cheap and reasonably good source of meat for piscivores. Remember that you should not feed it every day, especially when herbivores accompany the guapotes. Twice a week will suffice.

It is better to feed your fish twice daily than to give a double amount at one time. If food remains present five minutes after you have fed them, it was too much. It is not a bad idea to have some catfishes together with your cichlids. They scavenge the bottom and dispose of excess food. (Of course, the catfishes also have to be fed.)

Except tubifex, any other type of available live food can be offered to cichlids. Shrimp is the best fishfood there is, and it usually is available in frozen form. Krill and *Mysis*-like shrimp are excellent foods for cichlids. You could feed them twice daily. Before frozen food is given it must be completely thawed and rinsed under running cold water. During the freezing and thawing process tiny morsels of broken shrimp are present. These small particles are not noticed by fish and will put a heavy load on the metabolite contents of the water. When this happens you may see growth of blue-green algae in a previously stabilized tank. Hence the thawed food is rinsed before it is offered.

Cichlasoma breidohri female. Photo by Uwe Werner and Rainer Stawikowski.

FEEDING SPECIALIZATIONS IN CENTRAL AMERICAN CICHLIDS

Primitive predators from Caribbean islands

When the first cichlids ventured into Nuclear Central America some 25 to 50 million years ago, they encountered a rather deserted freshwater habitat. Nuclear Central America consisted mainly of Yucatan and northern Mexico. At that time cichlids were not the highly specialized fishes that we know today. It is most likely that those primitive cichlids lived on anything available. The main prey consisted of crustaceans and insects, and vegetable matter was probably eaten, as most primitive species still do today, but whether it was (is) of any nutritional value is not certain. Small fishes, however, were not disdained. In fact, the first cichlids could have wreaked havoc upon the small freshwater fishes that had populated Nuclear Central America (and the island bridge that connected the two Americas) from the north. In a habitat without important competitors the cichlids could thrive and grow to a relatively large size.

Central American cichlids were fortunate in that a stretch of ocean remained between Nuclear Central America and the mainland of South America. Not until two to three million years ago was the gap "repaired" by the connection of Panama and Costa Rica with Colombia. Until then Central American waters were dominated by secondary freshwater fishes and by so-called peripheral freshwater fishes. The latter group consists of species that thrive in fresh water as well as in sea water.

Although it is known that members of the family Poeciliidae were present from the beginning, it is evident that these small fishes could hardly be regarded as food competitors for cichlids. On the contrary, an explosive growth of these small livebearers would have been to the liking of many piscivorous cichlids.

Until two to eight million years ago, cichlids were in a relatively stable habitat and there were few or no forces or stresses to produce specialization. If specialization occurred it was due to the dense population of cichlids. The connection between Central America and South America allowed primary freshwater fishes to migrate into Central American rivers and compete with cichlids on the feeding level. Species from the family Characidae are very successful primary freshwater fishes, and their populations have outgrown those of the cichlids in Panama and Costa Rica. Further north characoids are not present in abundance; Costa Rica is as far north as most of them could reach in three million years. Together with the primary freshwater fishes, some South American cichlids also entered the waters of Panama (*Geophagus*) and Costa Rica (*Aequidens*). Central American *Cichlasoma* probably were derived from African species that lingered in the brackish waters of the northern coast of South America.

The present situation in the cichlid family of Central America is that the southern populations struggle against better adapted riverine species. This might be true, in some respect, of the northern populations as well. The central area (Honduras and Nicaragua) harbors the least disturbed and hence the most primitive species. Only in those areas where competition is low do primitive forms of cichlids exist.

Map showing distribution of the island-inhabiting cichlids of Central America. They were probably derived from African species which lingered in the brackish waters of the northern coast of South America.

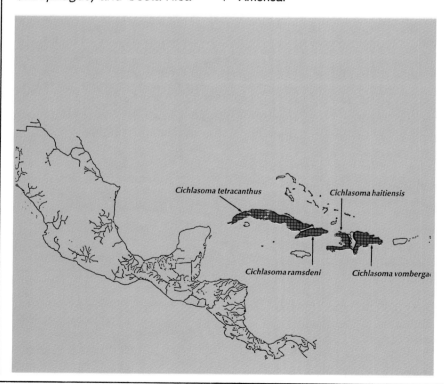

Cichlasoma tetracanthus

Cichlasoma haitiensis

Cichlasoma ramsdeni

Cichlasoma vombergae

How do we know what is a primitive form? *Cichlasoma woodringi* lived 25 million years ago and was found in the area that today is occupied by Honduras and Nicaragua. Over the millions of years both Nuclear Central America and South America drifted westward while the island (or shallows) between them (occupied by *C. woodringi*) more or less remained stationary. Now we call this island Hispaniola and it contains Haiti and the Dominican Republic.

In 1923 Woodring discovered on Haiti the fossil cichlid named after him. It was found that it had a close resemblance to present-day cichlids from Haiti and Cuba. Cockerell, who described the fossil, saw a very close relationship with *C. haitiensis* and *C. tetracanthus* and even placed *C. woodringi* in the subgenus *Parapetenia*! This subgenus (which should correctly be named *Nandopsis*) contains predatory cichlids whose prey usually consists of macro-invertebrates and small fishes. The remarkable fact is that the present-day cichlids from both islands (Cuba and Hispaniola) have hardly changed in their morphological features from features shared with the 25-million-year-old *C. woodringi*. The fact might be remarkable, but the reason is simple: due to the lack of competitors *C. haitiensis* and *C. tetracanthus* were not forced to specialize, hence these species are relatively unchanged from what they were millions of years ago.

As the mainlands drifted west, both Cuba and Hispaniola became isolated. It is very unlikely that cichlids crossed the Caribbean to reach them on later occasions. If they had it would have enhanced speciation on these islands or would have been reflected in the presence of part of the species flock that was developing on the mainland. None of this occurred.

In the last 10 to 20 million years Cuba and Hispaniola were always separated by at least 50 km (30 miles). Between them there is a stretch of rather deep Caribbean (Cayman Trough) that could not be crossed by cichlids. Still, *C. haitiensis*, from western Hispaniola, and *C. tetracanthus*, from Cuba, resemble each other to a great extent. If we realize that the complicated and overwhelming cichlid communities in the Great Lakes of Africa took much less time to develop from a few ancestors, it is clear that we might consider these Caribbean cichlids primitive.

Geographical variations may play an initial role in the development of new specializations. On the island of Cuba several geographical variants of *C. tetracanthus* do exist and were previously described as subspecies. Seen in the light of the variation shown in other Central American cichlids, it must be clear that the variability of *C. tetracanthus* does not merit splitting on the species level.

Another Cuban species is known: *C. ramsdeni*. In spite of the fact that this species has a very close relationship with *C. tetracanthus*, it seems to be a valid species. *C. ramsdeni* has a rather narrow distribution and is found only on the eastern side of the mountain ridge located at the far east of the island. It was probably well isolated from the populations of *C. tetracanthus* for a long period, during which it could have developed into a new species. Eastern and western Cuba were not united until some 10 million years ago. Successful geographical isolation often eventually leads to the formation of a new species. The same might be true for *C. vombergae*, a species with a similar origin as *C. ramsdeni*. *C. vombergae* was described from the Dominican Republic and was (is?) well isolated from *C. haitiensis* by the central highland of Hispaniola. Hispaniola originated from three separate islands that fused. The central highland forms the connection between the two main sections of Hispaniola. For an island population, some variants of *C. vombergae* show a remarkable degree of specialization. These variants possess over-developed lips similar to these of *C. labiatum* and may have specialized in feeding from rocky surfaces. Their diet may consist of mainly macro-invertebrates whereas the diets of the other island cichlids are of a more general nature.

Unfortunately, the original distribution pattern of the island cichlids is blurred by human intervention. On several occasions species from one region were transported to others and may have competed with the resident species. On Cuba, *C. ramsdeni* suffers the presence of *C. tetracanthus* but is especially threatened by the largemouth bass or trucha (*Micropterus salmoides*), a sunfish from Florida that was deliberately introduced into the waters of Cuba as a gamefish. The bass found the cichlid very edible, so much so that in order to rescue *C. ramsdeni* from extinction it was distributed throughout the island in 1939.

Of a different nature was the human intervention in the cichlid fauna of Puerto Rico. Thanks to the introduction of the predatory *Cichla ocellaris* and the highly competitive *Oreochromis mossambicus* and *Tilapia rendalli*, the original cichlid population completely vanished from this island. In 1862

Guenther described a cichlid from Barbados that was regarded as synonymous with *C. tetracanthus* by most ichthyologists after him. This species, originally named *Acara adspersa*, was living 1200 miles away from *C. tetracanthus*.

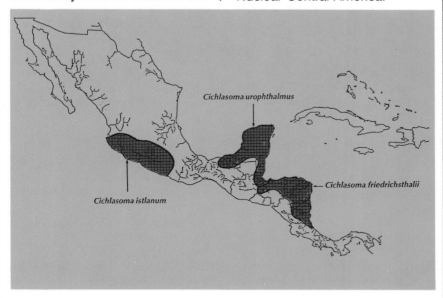

The colored portions of the map show the ranges of *Cichlasoma istlanum, C. urophthalmus* and *C. friedrichsthalii.*

This remarkable fact was doubted by most authors, and several supposed that this species was not caught on Barbados at all. On the other hand, it would not be surprising to find similar cichlids on the islands of the Antilles, since all of them once formed a bridge between Nuclear Central America and South America. By tectonic forces these islands were scattered over the Caribbean but may have retained their original cichlid populations. Maybe Guenther was right, and if so we may find in *C. adspersum* a key to the evolution of Central American cichlids.

Primitive piscivores from the mainland

Although island cichlids could prosper with a primitive feeding strategy, competition resulted in specialization on the mainland of Nuclear Central America.

Regarding cichlids, Central America can be divided into three biogeographical areas. This means that three different and virtually independent regions have developed three distinct lineages that constitute today's cichlid fauna on this sub-continent.

From north to south the regions are: Central Mexico, Yucatan/Guatemala, and Honduras/Nicaragua. These regions were isolated for a rather long period. Central Mexico was separated from Yucatan/Guatemala by the old Tehuantepec straits. The southern region, Honduras/Nicaragua, was virtually isolated by Amatique Bay. With some imagination we could reconstruct the evolution in these three regions and even explain the development of some specializations. Every region is thought to have been dominated by one ancestral species.

The "ancestral species" in the three regions are the most primitive and widely dispersed species known today. This does not imply that they really were the first cichlids in that area, but most likely the oldest that still remain. In my opinion, the "ancestral species" are for: Honduras/Nicaragua: *C. friedrichsthalii*; for Yucatan/Guatemala: *C. urophthalmus*; and for Central Mexico: *C. istlanum.*

The piscivorous cichlids from Central America all belong to the subgenus *Parapetenia* (recently *Nandopsis* has been found to be an older and preferred name), like the primitive species from the islands. It does not take a lot of imagination to see the piscivores as a specialization from the general feeders. Moreover, invading populations of small fishes were welcomed by these ancestors of the big guapotes, as they are commonly called. Competition split the dinner into an invertebrate part and a fishy part. The species of the subgenus *Amphilophus* are probably derived from the invertebrate-specialized *Parapetenia. Parapetenia* itself specialized to a piscivorous menu.

In general, only one piscivore could be expected in a single habitat, while selection would have favored the best adapted. Cichlids are very diverse, and among the piscivores of Central America several (sub)specializations have taken place. This is the reason why we may find *C. managuense* in the same waters as *C. dovii*; *C. salvini* and *C. urophthalmus* also have a widely overlapping distribution. *C. managuense* has specialized in small fishes from several different families while *C. dovii* is specialized to feed on other cichlids. A differentiation in

— *Cichlasoma managuense*

the feeding pattern of *C. salvini* and *C. urophthalmus* might also be present, and the latter grows twice as long as the first (30 and 16 cm respectively), but a previous isolation of the two geographically followed by recent expansion into the same areas may also explain today's overlapping distribution.

    C. managuense probably developed in the lakes of Nicaragua and is, as such, further specialized than *C. friedrichsthalii*; the species that may be the ancestor of *C. friedrichsthalii* has a rather wide distribution in Honduras and Nicaragua and grows to a "moderate" size of around 30 cm or 12 in (males). This species probably has the closest relationship to *C. tetracanthus* and *C. haitiensis* and has more or less a similar menu consisting of fishes and macro-invertebrates.

    Several races of *C. friedrichsthalii* are known. They originated from the geographical isolation of the streams and rivers in which this species occurs.

Long-term isolation usually leads to new species that will not recognize each other when brought together. The continental divide, which is present as a mountain ridge, acted as a complete barrier retarding the dispersal of several species. Due to the volcanic activities in the last 8 million years, several species were cut-off into two or more sub-populations. When the conditions (time and place) were favorable these isolated populations evolved into new variants and new species.

    In the distribution of most Central American cichlids we normally find the range restricted to either the Pacific or the Atlantic drainages (versants) of the continent. That this was not always the case is demonstrated by the fact that most species have their sibling species (barely distinguishable externally but acting like full species) on the other side of the divide. Once they were one and the same species dispersed over an area encompassing roughly the total

The colored area is the range of *Cichlasoma managuense*.

distribution of both species today. The rising of the continental divide separated the species into two populations that underwent independent evolution. This course of affairs is much more agreeable than to suppose an accidental "jump" over the continental divide and a subsequent spreading over the other side by the newly introduced fish.

    Resembling *C. friedrichsthalii* in behavior and distribution is *C. motaguense*. *C. motaguense* may have had a wider distribution along the Pacific drainage but has been successfully challenged there by *C. dovii*. *C. motaguense* grows to a slightly larger size than *C. friedrichsthalii* and reaches 35 cm (14 in) in length. These two species are so similar that it is not certain if the guapote found in the Rio Motagua really is *C. motaguense* or a race of *C. friedrichsthalii*. *C. motaguense*, a really beautiful species, is found mainly in El Salvador and Guatemala. Thanks to its excellent taste it was (and is) bred in large ponds. Introduction of the species into the waters of the Caribbean drainage in Guatemala may have happened for economic reasons several times in the past. This could explain the presence of this cichlid on both sides of the continental divide.

Lakes Nicaragua and Managua originated from volcanic activities prevalent in the area in the last 8 million years. In contrast to the African Great Lakes they are not completely isolated, nor do the surrounding rivers and streams contain water with a different chemistry. Nevertheless, these lakes proved to be optimum cichlid habitat, as are most lakes, and many new species developed. This speciation resulted from competition. When Lake Nicaragua was formed, the cichlid faunas from both the Pacific and the Atlantic drainages were united. The sibling species would have competed with each other for the same food, which was probably plentiful in the beginning. A high number of individuals increases the chance of producing a mutation that copes better with one particular type of food.

In today's lakes two piscivores are present: *C. dovii* and *C. managuense*. *C. dovii* is specialized to feed on small cichlids, while *C. managuense* relishes small fishes from other families, such as livebearers: The latter guapote could have developed from *C. friedrichsthalii*,

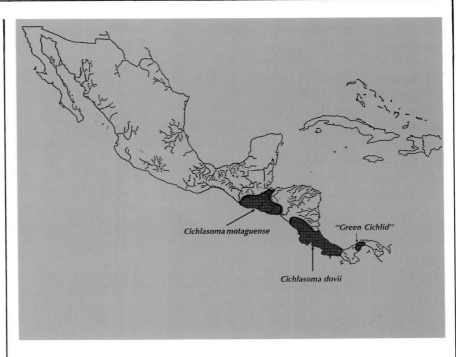

Cichlasoma motaguense

"Green Cichlid"

Cichlasoma dovii

The colored areas are the ranges of *Cichlasoma motaguense, C. dovii,* and the new fish, still unidentified and known as the "green cichlid."

and *C. dovii* from *C. motaguense* (or their ancestors). *C. friedrichsthalii* is still found in Lake Nicaragua, but since it cannot compete with its successor, the more specialized *C. managuense*, it is found only in river estuaries. *C. dovii* possibly was successful and may have driven its ancestor, *C. motaguense*, completely from the scene. Besides specialization in their feeding habits, both lacustrine piscivores are to be found in different habitats. *C. dovii* prefers a clear habitat with a rocky floor, but *C. managuense* favors a turbid environment over a muddy floor. These habitats are, of course, related to the type of food these species gather.

Specialized species stand a better chance in the competition with more primitive ones, especially with those from whom they were derived. Once *C. dovii* and *C. managuense* were specialized they moved out of the lake and dispersed over a larger area. Their natural spreading on the east side of the lake was hampered by mountains, but on

the Pacific side *C. dovii* dispersed into eastern Honduras. As an aggressive and large (up to 70 cm, 28 in, total length) predator, *C. dovii* may have conquered previous *C. motaguense* habitats. Due to its prized taste, *C. managuense* underwent an extensive and unnatural dispersal throughout Central America.

Further south, in Gatun Lake, Bleher and Mayland discovered in 1983 a large "Green Cichlid." This undescribed species might be a primitive predator from the Costa Rica/Panama section. However, the fact that it was found in a lake contradicts this idea. Further investigation of its distribution and behavior may give a conclusive answer to its position in the evolution of Central American cichlids.

It is fairly well known that many cichlids live longer and attain larger sizes than their

The spawning of *Cichlasoma dovii*. Photos by Rainer Stawikowski.

counterparts in heavily competitive ranges. This might especially apply to piscivorous predators like *C. dovii* and *C. friedrichsthalii*. These species have found a niche in the rich waters of Central America and hence aquired a correspondingly large size. The size of these piscivores, as is the size of any fish, is restricted by the amount of available food.

A fish is an ever-growing organism with a concomitant increase in the food intake. At the point the food supply cannot cover the needs of an individual, it immediately leads to the deterioration of the organism. The aged fish is weakened by the insufficient food supply and is eliminated from the population. The optimum length of any species is thus regulated by the amount of nutrition offered by the habitat. Under artificial conditions the life of the fish is clearly prolonged by the steady and sufficient supply of food and the lack of predators in its habitat. Prolonged life, clearly beyond its natural lifespan, may induce malformations like over-developed finnage, humped foreheads, and sometimes bulging contours.

From the Isthmus of Tehuantepec to Honduras, the Yucatan/Guatemala biogeographical region, a more or less isolated area harbored several cichlid species some three million years ago. The Atlantic drainage of this area was then separated from modern Honduras by Amatique Bay, but there was a connection on the Pacific side. The sea water of the bay may have prevented or hindered dispersal of the cichlid populations on both sides and may have led to an independent cichlid community in Mexico, Belize, and Guatemala.

The Rio Usumacinta was present as another bay in the same period and even may have formed a freshwater lake at a later stage. Such a lake would have contained similar but different species from both northern (Yucatan) and southern (Chiapas) areas. This would explain the diversity of the cichlid fauna found there today. Amatique Bay probably retained its marine nature, never yielding another specialization area for cichlids. The species from both areas (Mexico and Honduras) could only encounter each other at the margin, resulting in just a few new species. At the time the bottleneck at Amatique Bay allowed some *C. friedrichsthalii* ancestors to cross to the other side, a well-developed cast of predators was already present in Mexico. *C. friedrichsthalii*, which suffered hardly any competition in its own area, was confronted with more highly specialized predators in Mexico. At the same time that *C. friedrichsthalii* could pass through the Amatique bottleneck, other fishes also could pass. Among them the first characoids were present.

Before their passage, these characoids could have been dined on by *C. friedrichsthalii*, which could carry this habit into Mexico (along with its prey). Although several different predators were present in Mexico, none of them could have been specialized on characoids, because they were never there before. This opened the door for *C. friedrichsthalii*, which became an expert on these silvery delicacies. Since characoids move fast it had to adapt its feeding strategy. The extremely protrusible mouth of the predatory cichlid derived from *C. friedrichsthalii* allows for a powerful suction of the ambushed prey. In fact, the structure of the mouth is distinctive enough to place this specialized species in another genus—the species that possibly developed from *C. friedrichstahlii* is *Petenia splendida*.

Similar developments that had taken place in Honduras, El Salvador, and Nicaragua also occurred in Mexico. The most primitive piscivore here is probably *C. urophthalmus*. This species has a broad distribution but will not be encountered in great abundance in all of its range. It seems that this species is being successfully forced from its previous range. Only in those habitats that are avoided by the

---

Cichlasoma friedrichsthalii. **Photo by W. Heijns.**

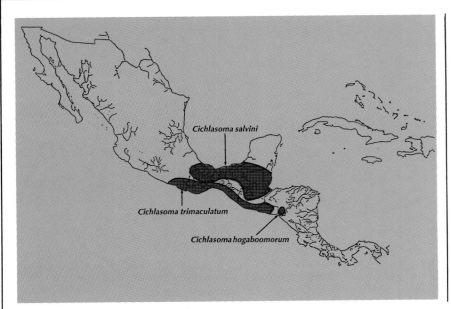

The colored areas of the map show the distribution of *Cichlasoma salvini, C. trimaculatum* and *C. hogaboomorum.*

resulted in new and specialized species. *C. salvini* could have been the result of such confrontations and could exist beside *C. urophthalmus*. Later, when eruptions threw up another mountain ridge, a population of *C. salvini* was isolated on the Pacific versant. This population could have evolved into *C. trimaculatum*, which is still found in the Pacific drainage of Mexico and Guatemala. It is probably not the only predatory cichlid in this habitat. At the edge of its range it may be encountered together with *C. motaguense*.

Although *C. salvini* and *C. urophthalmus* are from different

newer and more specialized species may we find *C. urophthalmus*. This means that *C. urophthalmus* is common in coastal areas and in other waters with a high mineral content. Due to its tolerance of saline water, this species could have spread through Honduras and Nicaragua via Amatique Bay. Thanks to its ancient traits even a distribution along both sides of the continental divide could have occurred. This would clarify the presence of *C. hogaboomorum* in the estuary of the Rio Choluteca in Honduras. This species resembles *C. urophthalmus* to a great extent, and it is also found in brackish water, albeit on the Pacific side.

In contrast to Amatique Bay, the paleo-Usumacinta "bay" was rather shallow. In fact, the whole region was close to sea level and an occasional flooding of extensive areas could have mixed several cichlid populations. This process has an effect similar to the formation of a lake. It

The yellow *Petenia splendida* above. Photo by Ad Konings. Below: A pair of *C. friedrichsthalii*. Photo by the author, Ad Konings.

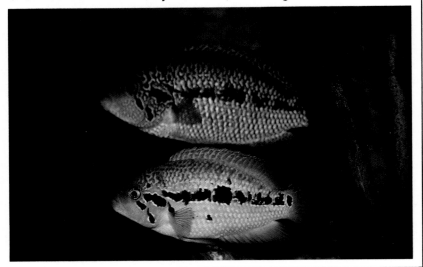

lineages, they are both predators with a generalized menu and might compete with each other.

C. salvini is much smaller than C. urophthalmus (16 and 25 cm, 6½ and 10 in, respectively) and may thus have distinct feeding habits. C. salvini is, however, better adapted to riverine conditions and has subsequently pushed many C. urophthalmus out of this habitat. On other occasions, as a rule, better adapted (for the same source of food) species usually are smaller. This is simple to understand: smaller individuals need less food and reach maturity sooner. The mutual effect increases the number of individuals per habitat without depleting the source of food. More individuals also means a greater chance for a successful mutation that will allow the species to compete at an even better level.

C. salvini and C. trimaculatus are normally found between weeds or roots and dine on a rather general type of food consisting of macro-invertebrates and small fishes. Another cichlid of the subgenus Parapetenia [Nandopsis] that can be found together with C. salvini is C. octofasciatum. Its scanty distribution might favor the idea that this species is not a well-specialized and successful cichlid. Its habitat includes swampy areas with warm, murky water. C. octofasciatum also may have known better days and may have thrived in the streams and rivers that harbor C. salvini today. Only the less favorable regions were left to C. octofasciatum. Unless C. octofasciatum can adapt itself to this uncommon habitat, it will be "chased" from this biotope as well. Only small numbers have been caught, and, although its iridescent color pattern is an adaptation toward the turbid environment, we might fear for its dim future.

During the last eight million years a frequent upturning of mountains and uplifting of highlands created a changing habitat for many cichlids in southern Mexico. When the Mesa Central was created by volcanism, it made a connection with the mainland to the northwest: the Isthmus of Tehuantepec was created. It is not known if the mainland (central Mexico today) already harbored cichlids. These cichlids could have been the predecessors of C. istlanum, C. beani, and the C. labridens group as well as those from the subgenus Herichthys. Since C. beani and C. istlanum are the only cichlids in their habitat, we might conclude that they are rather primitive compared to the specialized species from southern Mexico. It is possible that the cichlid present at that time could have resembled C. istlanum to a great extent.

When the predecessor of this predator or C. istlanum itself entered the southeastern area (Chiapas) after the formation of the Tehuantepec bridge, it became specialized into a true piscivore, but due to the great competition it had difficulty stabilizing itself in the right habitat. The uplifting of a large

Normally colored *Petenia splendida*. Photo by W. Heijns. Below: *Cichlasoma urophthalmus*. Photo by Werner & Stawikowski.

area in Chiapas isolated a population of these piscivorous cichlids in a new habitat that favored this species, and it still thrives today. The species is called *C. grammodes*. It is found in typical mountain streams with clear, oxygen-rich water. In such waters *C. grammodes* finds little food, so this species has developed a diet based on feeding on conspecifics. Outside this area *C. grammodes* or *C. istlanum* could not cope with the present species from the subgenus *Parapetenia*.

The original distribution of *C. istlanum* was broken up and developed into Pacific and Atlantic populations. In the Pacific drainage a further splitting resulting from the uplifting of the mountains in Jalisco and gave rise to two species: *C. istlanum* and *C. beani*. The latter is the most northerly cichlid on the Pacific side of the continent. On the Atlantic side there was apparently enough competition (e.g., from *C. urophthalmus*) to give rise to a species-complex centered around *C. labridens*. This complex probably originated from the predecessor of *C. minckleyi* and *C. pantostictum*.

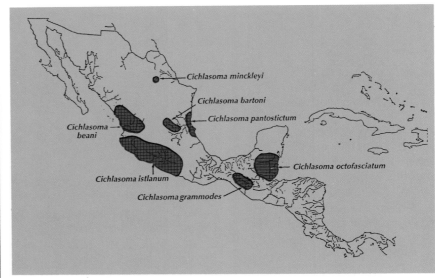

The colored areas on the map indicate the ranges of *Cichlasoma beani, C. istlanum, C. grammodes, C. minckleyi, C. bartoni, C. pantostictum* and *C. octofasciatum*.

The latter may have given rise to the species *C. labridens, C. bartoni,* and *C. steindachneri,* which are to be found in the Rio Panuco system.

C. pantostictum is restricted to coastal lagoons and other brackish waters. In this respect it resembles *C. urophthalmus*. Its descendants were better adapted to riverine conditions and might have pushed their ancestor out of the area.

C. labridens is dispersed over the main streams of the Rio Panuco system and is found in several races. Like the other species of this complex, it lives on a general assortment of food. Macro-invertebrates as well as small fishes top the menu of these *Parapetenia*. Competition with *C. carpinte* or similar species from the subgenus *Herichthys* has narrowed or altered the food-spectrum of some populations. These variants of *C. labridens* consume considerable amounts of snails, which they crush with their pharyngeal teeth. The snail's shell is also swallowed. In more advanced snail-crushers the shell fragments are expelled.

C. bartoni and *C. steindachneri* are restricted to small areas in the Panuco system and have adapted themselves to the clear waters of the mountain streams. The relatively poor food supply has transformed these once geographically isolated *Parapetenia* into specialized invertebrate feeders that may inhabit the same waters as does *C. labridens*.

Cichlasoma grammodes. Photo by Werner & Stawikowski.

Snail-crushers and leaf-choppers

The regions with minimum competition harbor the most primitive cichlids. This does not account for only island populations but also for those regions where life is harsh, such as regions with a high mineral content in the water or an unfriendly climate. When the circumstances are favorable, cichlids may thrive and abundantly populate the rivers and streams. When the number of cichlids increases, so does speciation. However, all these fishes are or once were in direct competition with each other.

The process of speciation is not so easy to understand, especially when we start the process with one species. It is evident and scientifically proven that two or more species can be derived from a single ancestor, but the processes involved always take the important and initial step of geographical isolation. Even in one body of water, two species can be geographically isolated. Example come to mind in the Great Lakes of Africa and even the Great Lakes of North America. Of course, mutations regularly occur among the individuals of one population, but most are lost in the process of unselective mating.

If the environment changes, for instance through the introduction of another species that is "dining from the same dish," a selective mechanism will favor those mutations that effectively tap another source of food. The consequence is that the old species is completely subdued by the newly developed species. There is no biological need to maintain a mixture of both "morphs."

A rich supply of different kinds of food may be eaten by a primitive cichlid. When there are no other fishes dining from this source there will be no selection toward a specialized form. Without cause, a specialization within the population (of one species) will not occur. Why should they specialize? None of the members of the group can be regarded as a better competitor. Only and solely from the presence of a competitor will feeding specialization take its course.

Evolution takes place due not only to changing physical environments but also by specialization. This is the reason why it may be so difficult to allot a certain population to a known species or to give it the status of a new one. The best example to demonstrate these problems is *C. minckleyi*. This species stands between the subgenera *Parapetenia* (= *Nandopsis*) and *Herichthys*. Species of the latter subgenus are more or less specialized to feed on vegetable material. The jaw teeth are closely set and may act like cutting blades. With this dental equipment members of this subgenus can chop leaves of higher plants. *C. minckleyi*, however, acts as a primitive predator, hence it belongs in the subgenus *Parapetenia*.

Variation is known among most Central American cichlids, but until recently it coincided with geographical isolation. Remember, for instance, the snail-crushing variants of *C. labridens*. The situation in *C. minckleyi* is different: here a mixture of typical snail-crushers and typical piscivores is found in the same water system. These two morphs recognize each other as conspecific and interbreed. This proves that these morphs belong to one species.

The problem arises when we consider the evolution of a new species. I just explained that specializations will never occur when there is no competition from

This underwater shot shows *Cichlasoma minckleyi* in their natural habitat. The darker fish is the male; the lighter fish is the female. Photo by Dr. I. Kornfield at El Mojarral.

A closeup of a *Cichlasoma minckleyi* female with her free-swimming fry. Photo by Dr. I. Kornfield in El Mojarral.

other species. Competition from *C. cyanoguttatum*, a species of the subgenus *Herichthys*, might have triggered the snail-crushing morph in *C. minckleyi*. The unexplainable fact is the co-existence of both morphs.

A changed habitat might also trigger speciation in a present cichlid population, but most geological processes occur gradually, preventing the sympatric existence of the unadapted and the specialized variants of one species. Volcanism may alter a cichlid's environment and force it to adapt. When adaptation, under such circumstances, is necessary, an extreme selection will wipe out the unadapted variant in a very short time. If the population was able to bring about an adapted variant, this variant would over-run the stressed and unadapted variant in a matter of years. The occurrence of such evolution can hardly be proven, but the fact still remains that a snail-crushing variant of *C. minckleyi* lives besides and breeds with an elongated variant of the same species. Their offspring contain

both forms plus a "hybrid" morph: an elongated body but with molar-shaped teeth on the pharyngeal apparatus.

The term "polymorphic species" means that the population of one species, throughout its total distribution, consists of two or more variants that vary in one or more inheritable features. This is true for every (!) species of living organism. Herein lies the basis of evolution. These polymorphic features, however, are usually undefinable and morphologically invisible. In some rare cases they can be easily recognized. An example is the "blue-tail" and "yellow-tail" morphs of some *Cyprichromis* species in Lake Tanganyika. Unequaled is the distinct polymorphic state of the pharyngeal apparatus of *C. minckleyi*. Although it is known that individuals of one population of *Aulonocara* in Lake Malawi have differently shaped pharyngeal tooth arrays, such an abrupt transition as is seen in *C. minckleyi* is unique among cichlids.

All anatomical features and

properties are inherited. The data from which the developing egg has to reconstruct the predetermined fish are stored on the chromosomes. These inheritance bodies determine every process during every step in the development (ontogeny) by the amino acid sequence called a gene. This you could compare with a file on a computer disk. When the file is needed it is read from the disk and the appropriate process is carried out. In most vertebrates chromosomes are present in duplicate (one set is derived from the mother, the other from the father), so each gene is present in pairs. This is a great advantage because when one gene (file) is defective, the complementary gene on the other chromosome will take over. In this binary system lies the basis for polymorphism. A gene can be defective and completely undecipherable but this inheritable file may also contain some additional data or some altered data that might still be read by the "ontogenetic computer." Such an altered gene may have given rise to the snail-crushing type in *C. minckleyi* and another gene to the body shape. This is of course a nice model in theory, but in practice you need a whole group of genes to result in the profound differences seen among the variants of *C. minckleyi*.

An indirect proof to this statement is the occurence of three to four intermediate types of pharyngeal teeth in one population of *Aulonocara* (e.g., *A. stuartgranti* from Mbenji Island).

Another possibilty that is offered in the literature is the frequent sparsity of food in the natural habitat of this enigmatic cichlid (*C. minckleyi*). It is scientifically shown that the snail-crushing morph switches to hard-shelled snails when the food supply is at minimum. The elongated type with the slender pharyngeal teeth has to succumb during feeding stress. If this were true, it would be nonsense and spilling energy to retain the elongated variant. During food scarcity, most if not all of the latter variant would die out.

It was also shown that the variety and quantity of food intake did not differ among the variants of *C. minckleyi*.

Personally, I think that the problem with *C. minckleyi* can still be explained. As I argued before, geographic isolation should be involved. Two populations of *C. minckleyi* were isolated for a long period of time (perhaps 100 years, perhaps one million years), the separation being long enough to develop adaptations toward their respective environments. The separation, however, was not long enough to form a new species, because when these two population were again brought together they still recognized each other as conspecific.

We have discussed geological changes that split up a population and eventually led to the formation of another species. Why is it unlikely to think about the reverse of the process? *C. minckleyi* may have had a wider distribution, but due to the changed climate (it became warmer) some parts of its habitat dried up. This might have divided the previous population into two. After a long period the water level may have risen again and may have united the two populations. The natural habitat of *C. minckleyi* lies around Cuatro Cienegas in northern Mexico. These waters are drained by the Salados River into the Rio Grande. A possible rising of the water level could have occurred when the river was dammed downstream near Laredo and the Presa Venustiano Carranza was created.

During the period that *C. minckleyi* was separated into two populations, a whole set of new genes (files) could have developed in the snail-crushing variant. These new genes are now mixed with the unspecialized ones. As long as these genes (files) are located on one or two chromosomes (diskettes) they will be expressed as true morphological characters without transitory morphs, and as long as the food supply is sufficient these two morphs may exist for quite some time sympatrically. Later, slow processes such as chromosome breakage and exchange may smooth out the morphological contrast that is evident in the modern population. In this context we should not forget *C. cyanoguttatum*, which is found in the same waters and feeds mainly on vegetable material. Since the original and elongated *C. minckleyi* is a primitive predator, it means that *C. cyanoguttatum* poses some competition. This may finally lead to selection in favor of the snail-crushing morph.

C. minckleyi nicely demonstrates a failure in speciation because the geographical isolation of the two subpopulations did not last long enough.

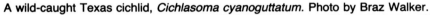

A wild-caught Texas cichlid, *Cichlasoma cyanoguttatum*. Photo by Braz Walker.

Above: A female *Cichlasoma cyanoguttatum* collected and photographed by Dr. Herbert R. Axelrod in Texas. Below: The same fish spawning.

Primitive vegetarians

There is hardly any cichlid that is strictly herbivorous, although some species in the Great Lakes of Africa are 99% vegetarian. Such highly specialized forms are not encountered in Central America. The development of a herbivore among cichlids is not an easy process. True, the oldest and least specialized species (e.g., *C. tetracanthus*) partially live on vegetable matter, but this is not an important item in the diet. The real herbivores need a considerable adaptation concerning their intestines. Vegetable material is difficult to process, and only a small percentage of the bulk will be digested. This implies an enlarged volume of the intestine and a steady supply of food. Several traits in this direction have been developed in Central America, and a few species are highly specialized for digesting vegetable material.

Regarding the evolution toward an herbivorous diet, we might again divide Central America into the three major regions. The most primitive herbivores are found in northern Mexico and belong to the biogeographic region of Central Mexico. These species belong to the subgenus *Herichthys* and are probably derived from *Parapetenia* (= *Nandopsis*). Typical for primitive species is the variation among several populations of the same species. In areas where there is little competition we may find unspecialized herbivorous species, but when competition is high we encounter snail-crushers or strict herbivores. This is true for *C. cyanoguttatum*, the most northerly cichlid. Its distribution is centered around the Rio Grande del Norte and it is found in Texas, hence its common name Texas cichlid.

For a long time *C. carpinte* was

The colored areas indicate the ranges.

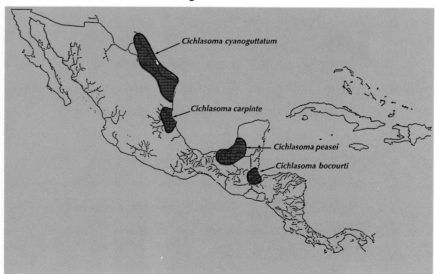

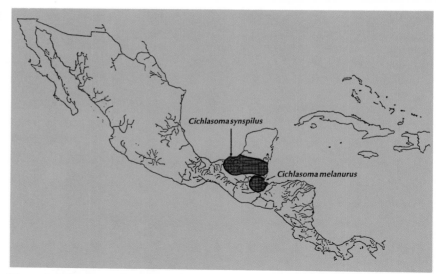

The colored areas indicate the ranges.

considered a synonym of *C. cyanoguttatum*, but its distribution clearly lies in Mexico, in the drainage of the Rio Panuco. Here it is found together with cichlids from the *C. labridens* group. The teeth on the jaws of *Herichthys* species are tightly set and form a cutting edge that can chop leaves from plants, aquatic as well as terrestrial. Maceration of the vegetable material also may be performed by the grinding process of the pharyngeal teeth. This may lead to strong teeth on the pharyngeal jaw apparatus, sometimes misinterpreted as belonging to a snail-crusher. *C. cyanoguttatum* and *C. carpinte* are primitive herbivorous species with numerous variants that are expressed in morphology or in coloration. The typical variability of primitive species hampers the correct classification of the variants. Despite their taxonomic problems, these cichlids are very popular and easy to maintain in captivity. The maximum size lies around 25 cm (10 in), which is not too big for a 100-gallon tank.

Recently some new species or variants of *Herichthys* were discovered in Mexico. A distinct species was found in the Rio Nautla, north of Vera Cruz. The richly colored representatives of this species reveal their ancestry only when in breeding color.

Once the species of the subgenus *Herichthys* must have had a much broader and denser distribution. When the Isthmus of Tehuantepec rose above sea level, *Herichthys* was able to spread throughout the lower biogeographic region, Yucatan/Guatemala, as well. The concomitant evolution of herbivorous species from the subgenus *Theraps* resulted in a limited habitat for *Herichthys*. At the moment the Isthmus of Tehuantepec could be crossed, the herbivorous *Theraps* were

probably not specialized enough to prevent dispersal of *Herichthys* throughout this biogeographic region. The presence of *C. bocourti*, a member of the subgenus *Herichthys*, in Guatemala and in southern Belize favors this idea. *C. bocourti* might be considered primitive in respect to *C. pearsei*, another *Herichthys* that is found together with species from the subgenus *Theraps*. Due to the pressure of the herbivorous *Theraps*, *C. pearsei* specialized in feeding from higher plants, aquatic as well as terrestrial. When the water level is high *C. pearsei* eats the grass and the weeds from the flooded land.

A few million years ago the cichlids of Yucatan/Guatemala were confronted with herbivores from the northern region. This triggered the development of more specialized herbivores in the local populations. The predecessor of the herbivorous *Theraps* in Mexico was probably a primitive species feeding on a general menu, including vegetable matter. It could have been a member of the early *Parapetenia*, which also are known to feed on vegetable

debris. The most important adaptation of the herbivorous *Theraps* in this region is the enlargement of the intestinal tract. This allows them to collect large quantities of food that will be processed gradually. These herbivores feed on vegetable debris such as half-rotten leaves, algae, aquatic plants, seeds, and fruits. Insects and crustaceans, however, are not disdained, completing the diet of these weakly specialized herbivores.

Interestingly for the hobbyist, this group of cichlids contains the most beautiful species you might come across in Central America: *C. fenestratum, C. synspilus, C. melanurus, C. guttulatum, C. bifasciatum, C. regani, C. hartwegi*, and the recently described *C. breidohri*. When you look at these species it must be clear that all of them have a common ancestor. In fact, no two of these species (with one exception) have an overlapping distribution, and they thus are derived by geographic isolation. Only *C. synspilus* and *C. melanurus* are found sympatrically with *C. bifasciatum*. *C. synspilus* and *C. melanurus* are never found together and are very closely related. These two species are specialized to feed on fruits that fall from the trees above their habitats. *C. bifasciatum* is more of a generalized herbivore and might co-exist with the other two. The common ancestor might well be

large rivers and especially in the brackish water of the estuaries. The tolerance of this species toward a high mineral content in the water was of great advantage to the cichlid: it could spread along the complete Atlantic coast of Central America and become

the cichlid with the widest distribution in this part of the world. *C. maculicauda* is found from Tehuantepec to the Panama Canal.

C. hartwegi has a similar origin as described for *C. grammodes*. *C. hartwegi* resembles *C. bifasciatum* to a great extent and is probably a direct descendant of this species. A population of *C. bifasciatum* in the Rio Grande de Chiapa was isolated from the remaining lowland population when the region rose to 2000 meters. *C. hartwegi* is the result of this physical separation and is found together with *C. grammodes*, a specialized piscivore, in the rivers around Presa de la Angustora. In the Presa de la Angustora, a lake formed by the rising downstream of the Rio Grande de Chiapa, another variant or species may have developed: *C. breidohri*. This species could as well be a variant of *C. hartwegi* or a derived species specialized for feeding on insects and crustaceans. According to the discoverers, Stawikowski and Werner, this

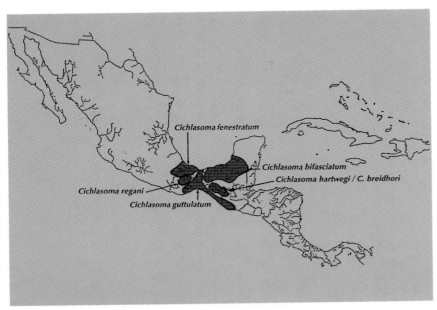

The colored areas indicate the ranges.

C. maculicauda (or *C. heterospilus*), which has a close relation with all of these herbivorous *Theraps*. *C. maculicauda* originated in the region Yucatan/Guatemala and dispersed by sea over the whole of Central America. From *C. maculicauda*, a type like *C. synspilus* and a type like *C. bifasciatum* could have evolved. Since the latter two are better specialized, they could be thought of as having pushed *C. maculicauda* to the margins of the habitat. In fact, the species was not pushed there but was left in peace, since these habitats were not preferred by the more highly specialized herbivores. This explains why *C. maculicauda* is normally found in the middle of

Cichlasoma bifasciatum. Photo by Dr. Harry Grier, courtesy of the Florida Tropical Fish Farms Association.

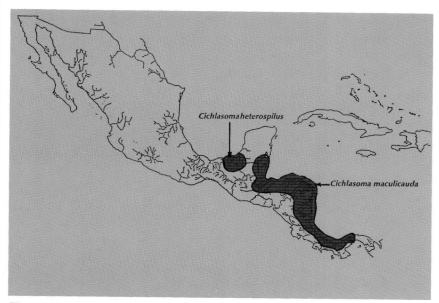

The colored areas indicate the ranges.

species was found sympatrically with *C. hartwegi*. Such sympatry would indicate that these two cichlids are not conspecific.

Specialized herbivores

Theraps is a large subgenus including several groups. The group that we discussed before, I called herbivorous *Theraps*. This is in distinction from the insectivorous *Theraps*, which include snail-feeders as well. Originally *C. irregulare*, an invertebrate-feeder, was the first cichlid to be described in the section *Theraps*. In the near future a revision of the Central American cichlids will certainly redistribute and modify the available groups in the respective subgenera. Since several groups are to be discriminated among *Theraps*, this name will be applied to the group of invertebrate-feeders of the current subgenus. The herbivorous *Theraps* will be placed in another genus or subgenus, but not in the recently described *Paratheraps* Werner & Stawikowski, 1987. The authors of the latter description failed to assign a type species to the

genus, which makes the generic name invalid (Geerts, 1988).

The next two species still belong to the subgenus *Theraps* but are specialized herbivores. *C. intermedium* and *C. godmanni* are distributed over the Yucatan/Guatemala region and are found in the same area as some other vegetarians from the same subgenus. This calls for an explanation, as species from the

same (sub)genus with similar feeding behaviors usually are not found together. The reason is that *C. intermedium* and *C. godmanni* normally are found in clear waters and feed rather specifically on fluffy algae from rocky substrates. The whole group of taxa centered around *C. bifasciatum* is more often encountered in turbid waters and feeds on less specific vegetable matter. *C. godmanni* is distributed around Amatique Bay (Bahia de Amatique) and may have dispersed through great parts of Honduras. The cichlid fauna of Honduras and part of Nicaragua has not been extensively investigated and might still hide some herbivorous *Theraps*, probably not plant-eating *Theraps* but the cave-breeding, insectivorous *Theraps* found further south over Honduras and Nicaragua.

A prominent characteristic that sets this group apart from the herbivorous *Theraps* is the fact that they seek refuge among the jumbled rubble of the bottom and breed there, whereas the larger herbivores are usually seen over soft bottoms and employ the

The colored areas indicate the ranges.

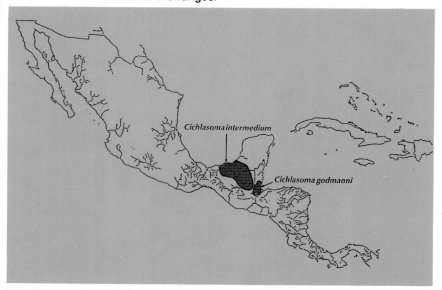

open substrate breeding method. The *Theraps* encountered south of their origin in Yucatan/Guatemala belong to the cave-breeding type. Hence it is more likely that the southern *Theraps* were derived from the slender and elongated insectivorous *Theraps*.

A rather unspecialized species of this group is *C. microphthalmus*. The ancestor of the southern *Theraps* probably was similar to this species from Guatemala. It would not be surprising if more *Theraps* species were found in Honduras and Nicaragua. They would be the missing links in the seemingly discontinuous dispersal of this subgenus.

On the other hand, it seems possible that the riverine *Theraps* displaced the local insectivores (*Amphilophus*) in this part of Central America. This could be partially due to the lack of appropriate habitat, like rocky bottoms, or due to the simple fact that *Theraps* was, at that time, not the specialized insectivore it is today. It is, however, likely that such species as *C. nicaraguense*, *Neetroplus nematopus*, *C. tuba*, *C. sieboldii*, and *C. panamense* were derived from a common ancestor that might have resembled *C. microphthalmus*.

When Lake Nicaragua was created several cichlid species were confronted with other species that had similar feeding habits. As long as food was abundant, there was no need to change to secure the future of the species. Cichlids have the ability to adapt to changing environments or scanty food supply. As soon as food became scarce the specialization process was enhanced. Only those mutants that could utilize an unused or hardly used food source were favored. Due to feeding specializations the lake

became densely populated and many cichlids had difficulties in acquiring suitable spawning sites. This triggered the evolution of breeding mechanisms.

The local insectivores of the subgenus *Amphilophus* could have stressed the immigrant *Theraps* and pushed them toward specialization. One new species became a snail-feeder (*C. nicaraguense*) and another new species specialized on algae (*Neetroplus nematopus*). The ancestor of these species also spread more to the south and founded populations in Costa

Rica and Panama (*C. tuba*, *C. sieboldii*, and *C. panamense*).

C. panamense is the most highly specialized *Theraps* outside Lake Nicaragua that developed from the southern branch of this subgenus. Its teeth are adapted to scrape not only algae but also invertebrates from rocky surfaces. At one time it was placed in the genus *Neetroplus* since its teeth resemble those of the single species in this genus (*N. nematopus*). Later it was found that the chisel-shaped teeth of *C. panamense* developed during maturation and could as

The colored areas indicate the ranges.

Cichlasoma trimaculatum spawning. The female is shown depositing her eggs in the open. Photo by Uwe Werner.

well be an acquired feature instead of an inherited one (Rivero, 1981). This brought *C. panamense* back into the large genus *Cichlasoma*, but a close relation with *N. nematopus* could not be overlooked. Fry and juveniles of the species *C. tuba* and *C. sieboldii* resemble those of *N. nematopus*, which corroborates the idea that these two species also have a close kinship with that highly specialized algae-scraper.

A hypothetical reconstruction of the evolution of the southern *Theraps* might be as follows. A rather unspecialized (*C. microphthalmus*-like) insectivorous member of the subgenus *Theraps* could have spread through Honduras and Nicaragua but was halted by the size of Central America in those times. When Lake Nicaragua was formed, several species found themselves in the newly created lake and started their interaction. With time several specializations evolved, and for some species the new lake formed a convenient bridge to the south and they dispersed over Costa Rica and Panama.

By the time *Theraps* migrated into Costa Rica the species involved were descendants from the herbivorous fishes that originated in Lake Nicaragua or at least in the streams of Nicaragua. Further dispersal followed, and the later *C. panamense* settled in Panama. The continental divide separated the Costa Rican population into *C. tuba* (Atlantic) and *C. sieboldii* (Pacific). *C. panamense* is reported from both drainages but may be divided into two (sub)species. This is indicated by the fact that two distinct color morphs are known from this cichlid, although their precise distribution is unclear. The construction of the Panama Canal bridged the two drainage regions in Panama and may have obscured the original distribution pattern.

C. tuba may grow to a size of about 30 cm (12 in). A considerable part of its diet consists of fruits. Some individuals have overdeveloped lips that resemble those of *C.*

Cichlasoma labiatum. Photo by Mervin F. Roberts.

This fish is from Africa! It is probably *Cyrtocara euchila,* but it does demonstrate the labial enlargement. Photo by Dr. Warren E. Burgess.

labiatum and *C. vombergae,* but this is probably not an adaptation for the fruit-eating diet. *C. tuba* might not have given up its previous feeding habits—some populations might have specialized in sucking insects from rocky substrates. Several African cichlid species from different lineages (e.g., *Cyrtocara euchila, Lobochilotes labiatus*) possess overdeveloped lips as a specialization toward a particular

feeding behavior. In all cases the fleshy lips act like gaskets that seal off a small hole and enable the fish to suck out the prey. The swelling of the lips also could be induced by the rough substrate that is foraged on by the fish. Although *C. tuba* is known to eat vegetable material, it is probably less advanced in doing so than *C. panamense.* This could be the reason that its distribution is more restricted than that of its sibling species, *C. sieboldii.*

From South America (Colombia and Ecuador) a similar sized cichlid, *Aequidens coeruleopunctatus,* migrated into Central America. This cichlid feeds on invertebrates and, although it prefers another habitat, it might have pushed *C. tuba* to its present distribution in Costa Rica. *C. tuba* and *C. sieboldii* are normally encountered in clear waters with a rocky bottom. Both are reported to be cave-breeding species, although *C. tuba* may spawn in the open like *A. coeruleopunctatus.*

C. sieboldii grows to a maximum size of about 20 cm (8 in) and could have a differentiated diet in respect to *C. tuba* and *A. coeruleopunctatus.* It is a real cave-spawning species and is encountered in clear, rock-strewn streams. A herbivorous diet is not reported but can be anticipated from its ancestry.

The most specialized descendant from the original *Theraps* species is *Neetroplus nematopus.* In confrontation with other species, *N. nematopus* evolved into a successful cichlid that inhabits all rocky areas inside Lake Nicaragua. This rock-bound cichlid is the Central American equivalent of an African mbuna and is one of the most abundant cichlids in the lake. Its chisel-shaped teeth have a striking similarity to those of *Eretmodus cyanostictus* from Lake Tanganyika. They are excellent instruments with which to scrape algae from rocks. Not only algae but also zooplankton form the favorite dishes of this fish. In this respect it is a perfect homolog to the mbunas. Its successful feeding adaptations, which undoubtly developed in the lake, brought *N. nematopus* beyond its place of birth. It is recorded from the total Atlantic drainage of Nicaragua and northern Costa Rica. Outside the lake, *N. nematopus* prefers clear waters with rock-strewn bottoms. The riverine populations show a more elongated body profile that seems to be an adaptation toward running water. It is one of the smallest Central American cichlids, but despite its maximum size of only 15 cm (6 in) it is able to defend a territory effectively. Both male and female will attack sometimes much larger intruders and with their razor-sharp teeth may produce serious wounds. An interesting step in cichlid evolution was the development of a herbivore (*Herotilapia multispinosa*) from an insectivore (*C. centrarchus*). We have seen how plant-feeders have developed from *Parapetenia* in Central Mexico (*Herichthys*) and in Yucatan/Guatemala (*Theraps*), and in Honduras/Nicaragua a herbivore also was created. Several biotopes and food resources were explored by this group of Central American *Parapetenia*. *Parapetenia* stood at the base of all variants that developed in these regions. The generally large size of these primitive predators was unsuitable to populate the shallow waters. Hence a group of small *Parapetenia* split off and underwent evolution into the modern subgenus *Archocentrus*. Because of their size they could not be regarded as notorious

Cyrtocara euchila from Africa. Photo by Dr. Herbert R. Axelrod.

Another form of *Cyrtocara euchila* with the egg spots in the dorsal and not the anal. Photo by Dr. Warren E. Burgess. Below: *Lobochilotes labiatus* from Lake Tanganyika, Africa, where it was collected and photographed by Dr. Herbert R. Axelrod.

A closeup of the enlarged lips of *Cyrtocara euchila*. Note the rust-colored teeth imbedded in the lips. Photo by Dr. Herbert R. Axelrod.

piscivores, but they were respected as insectivores. Shallow water may flow over different bottoms and the small *Archocentrus* species therefore developed in two directions. Most of them chose the rocky habitat. *C.(A.) centrarchus* preferred soft bottoms with submerged vegetation. The vegetation in its environment also allowed *C. centrarchus* to feed on plants and algae.

Due to the competition from another species (probably juvenile *C. managuense*) a population of *C. centrarchus* specialized in feeding exclusively on vegetable matter. This specialized variant further adapted itself to the circumstances and developed a new type of tooth. Each tooth comprises three formerly separate teeth that have fused to form a cutting edge. Since cichlids are classified according to their dentition, this peculiar descendant from *C. centrarchus* was placed in another genus, *Herotilapia*.

The species I am talking about is *Herotilapia multispinosa*. Its deviant dentition is a highly specialized feature, and its development was enhanced by the fact that *H. multispinosa* purposely isolates itself in small ponds and backwaters created in the dry season when the water level drops. These very shallow pools are inhabited by species of the family Poeciliidae and by *H. multispinosa* and juvenile *C. managuense*. The latter, an invertebrate-feeder at this stage, may have triggered the development of the insectivorous *C. centrarchus* into the algae-eating *Herotilapia*. Since *C. managuense* had its start in Lake Nicaragua, *Herotilapia* also may have developed along the banks of this lake. Due to the

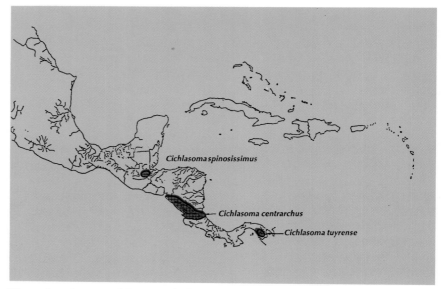

The colored areas of the map indicate the ranges.

Herotilapia multispinosa tending its fry, which have been hung on the bush. Look closely. Photo by Hans Joachim Richter.

shallowness of the pools the temperature of the stagnant water may soar into the low 30's C (high 80's F). Growth of the vegetation is thus enhanced and a thick blanket of fluffy algae is spread out to be eaten by *H. multispinosa*.

Another great advantage for this species is the fact that they breed in these pools too.

The presence of a first-class algae-scraper (*N. nematopus*) in the rocky sections of the Nicaraguan rivers and an optimally adapted *Herotilapia* in the stagnant and weedy parts made life difficult for the other herbivores (if any exsisted in addition to *C. maculicauda*). Both species are, however, restricted to the shallow sections, which

leaves room in deeper water for a few large herbivorous species of the subgenus *Theraps*, which is, however, only represented by *C. maculicauda*.

In Panama, a large herbivore is known from the Pacific drainage: *C. tuyrense*. Like the large herbivorous *Theraps* in Yucatan/Guatemala, this species lives in turbid water and feeds on algae and higher plants. Even terrestrial plants are eaten when the swollen rivers flood the land. Although there is a considerable resemblance to the northern herbivorous *Theraps*, *C. tuyrense* has its roots in the subgenus *Amphilophus*. This subgenus consists of detritus-feeding species usually occurring over soft bottoms. The competition of

Aequidens coeruleopunctatus and *Geophagus crassilabris* in Panama forced the ancestor of *C. altifrons*, a general bottom-feeder, to specialize toward a more restricted diet. The result could have been two-fold: in addition to herbivorous *C. tuyrense*, a more piscivorous species could have developed, namely *C. calobrense*. Both are to be found in the same drainage, but *C. calobrense* has a wider distribution. *Herotilapia multispinosa* dispersed throughout a great part of Central America but is not yet reported from Panama. Nothing is halting this algae-eater from populating Panama as well; only human intervention can wipe out such a successful species.

A pair of the yellow variety of *Herotilapia multispinosa* spawning. Photo by Ruda Zukal.

The female *Herotilapia multispinosa* changes color as her fry become free-swimming and begin to venture from the bush. Photo by Hans Joachim Richter.

A juvenile specimen of *Cichlasoma intermedium* (?). Photo by Werner & Stawikowski.

Invertebrate-feeders from the subgenus *Theraps*

The primitive cichlids of the subgenus *Parapetenia* were dispersed over all three biogeographical regions discussed before. The northern region, Central Mexico, is inhabited by somewhat primitive cichlids. Most of them are not specialized at all, but two groups can be discerned. One group, *Herichthys*, has put emphasis on vegetable matter, while the other group, *Parapetenia*, is more interested in fish and invertebrates. The lack of a pronounced confrontation with other species with a similar feeding behavior resulted in a cichlid community with marginal specializations. The rather primitive state of these cichlids yielded the greatest variety among species found in Central America. If the competitive stress would increase, these variants might easily become true species and become even more adapted than their variability, within the species, allowed them until now. The few true invertebrate-feeders found in this region are undoubtedly immigrants from the Yucatan/Guatemala area.

Besides an herbivorous trait, an insectivorous branch developed from the initial *Parapetenia* in Yucatan/Guatemala. The herbivorous *Theraps* from this region have been discussed and were found to contain three subgroups: algae-feeders (e.g., *C. intermedium*), fruit-eaters (e.g., *C. synspilus*), and general herbivores (e.g., *C. guttulatum*). The invertebrate-eating *Theraps* developed into four subgroups: general insectivores (e.g., *C. microphthalmus*), insectivores from clear waters with rock-strewn bottoms (e.g., *C. lentiginosum*), snail-feeders from turbid waters with soft bottoms (*C. heterospilus*), and algae-scrapers from clear waters with rocky bottoms (e.g., *C. bulleri*). In Yucatan/Guatemala a glittering array of fine cichlids suitable for the aquarium are invertebrate feeders of the subgenus *Theraps*. These species reach an average length of about 20 cm (8 in) and are not easy species to breed in captivity. These facts make those beautiful species very interesting to maintain and observe in the confined surroundings of a tank.

Clear waters usually lack an abundant food supply, hence they are not as densely populated as are turbid streams. The reason is simple. Most clear waters are continuously running and thus inhibit the precipitation of sediment on the normally rocky bottom. Sediment forms the basis of rich plant growth, including algae and aquatic weeds. Running waters also have a generally lower temperature than do stagnant ponds and lakes. Thus dense plant growth is further hampered by a lower temperature in respect to the temperature downstream. Where the stream widens the current slows down and sediment may accumulate. Moreover, the temperature here rises and enhances plant growth. The induced algal growth plus the washed-down sediment usually fill broad rivers with turbid but rich water. It is here that the herbivorous *Theraps* are found, not upstream.

A snail- and invertebrate-eating cichlid did adapt itself to this environment: *C. heterospilus* of the Rio Usumacinta drainage system. This species probably has the closest relation to the herbivorous *Theraps* (it has a great resemblance to *C. maculicauda*). *C. bifasciatum* is found in the same waters and may have the same ancestor as _C. heterospilus._

A less specialized cichlid is _C. microphthalmus._ This general invertebrate-feeder inhabits the Atlantic drainage of Guatemala and is found over soft as well as over hard bottoms. This species (or a similar ancestor) may have spread _Theraps_ further south, giving rise to snail-feeders and algae-scrapers.

C. irregulare, the type species of _Theraps,_ is encountered in the same streams as is _C. microphthalmus. C. irregulare_ is, however, adapted to clear water and rocky bottoms. As such it may successfully compete with the other insectivores found in the same drainage system. _C. irregulare_ has very close affinities to _C. lentiginosum, C. coeruleus,_ and _C. nebuliferum._ These four species have a common ancestor and were created through geographical isolation. Typical of all four species is the rocky habitat and clear water in which they live. The rocks give them shelter and spawning sites. These species are cave-breeders.

Cichlasoma bifasciatum or a female _synspilum._ **Photo by Werner & Stawikowski. Below: The colored areas indicate the range of the species.**

From these rock-bound invertebrate-feeders a new specialization evolved: scraping biocover from the rocks. The three species involved, _C. bulleri, C. gibbiceps,_ and the recently described _C. omonti,_ might be compared with _Labeotropheus_ from Lake Malawi. _Neetroplus nematopus,_ a southern descendant from _Theraps,_ resembles more closely _Pseudotropheus_ from the same African lake. The similarities are especially related to the anatomical features. The diet of _C. bulleri_ and _C. omonti_ contains a considerable amount of algae but invertebrates also form an important percentage.

Whereas the species from the group centered around _C. lentiginosum_ deliberately pick their prey from the biocover, _C. bulleri, C. gibbiceps,_ and _C. omonti_ might be involved in a rather continuous scraping. The random food intake of the latter three should include a large

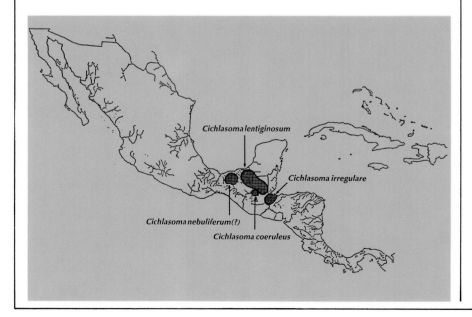

Cichlasoma lentiginosum

Cichlasoma irregulare

Cichlasoma nebuliferum(?)

Cichlasoma coeruleus

Cichlasoma bulleri. Note the snout and inferior position of the mouth. Photo by Werner & Stawikowski.

The map below indicates the ranges by the colored areas.

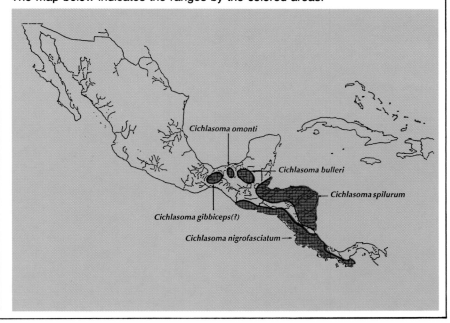

amount of algae together with an occasional insect or crustacean that was concealed in the biocover. The occurrence of invertebrates in the biocover is likely to be rather frequent because their usual shelter, soft bottoms, is lacking in this particular habitat. In fact, a rocky section in a river may harbor a completely different set of cichlids than a soft-bottomed part of the same system. Both habitats may contain insectivores, but it may not always be the same species.

The mutual ancestor of *C. nicaraguense* and *N. nematopus* was a cave-breeding species from the insectivorous *Theraps* lineage, and it was found in a rocky habitat. The presence of another small insectivore from the subgenus *Archocentrus* forced this ancestor to develop other specializations or be replaced by *Archocentrus*.

The insectivorous *Theraps* evolved into two new species. One, *N. nematopus*, became an aggressive scraper of the biocover; the other species, *C. nicaraguense*, tried its luck on the lake floor. The latter specialized on snails, feeding from the sandy floor. The successful adaptation of *C. nicaraguense* made it the most common cichlid from the lake.

C. nicaraguense originally was an invertebrate picker and may have selectively picked its food (mainly snails) from the sandy floor. Despite the small size of the mouth, it developed into an underslung position. *C. nicaraguense* "chews" the sand in search for something edible. This cichlid kept its previous habit of breeding in caves, hence it has to find a suitable place among the territories of the resident cichlids of the rocky coasts.

Small predators from the subgenus *Archocentrus*

In order not to complicate the evolutionary description too much, only two herbivores will be discussed in this section. The reason is that they are undoubtedly derived from insectivorous members of the subgenus *Archocentrus*. The evolution of *Archocentrus* has taken place in the Honduras/Nicaragua region.

Archocentrus is a direct descendant of *Parapetenia*. The small species of the newly developed subgenus could populate the shallow waters where a wealth of food was

awaiting predation by cichlids. The various members of the subgenus are selective pickers, hence they may prefer clear water and a rocky bottom.

One of the first *Archocentrus* to evolve was *C. nigrofasciatum*. The success of this relatively small 15 cm (6 in) cichlid is demonstrated by its wide dispersal over the Pacific drainage of Central America from Guatemala into Panama. Through Lake Nicaragua, *C. nigrofasciatum* could expand its distribution over the Atlantic side of Costa Rica. The Rio San Juan drains from the lake and facilitated the dispersal of most lacustrine species over its drainage area. In some regions *C. nigrofasciatum* is even found over soft bottoms, but its preferred habitat is the clear and shallow water of rock-strewn mountain streams.

All species of the subgenus *Archocentrus* have more spines in the dorsal and anal fins than do other Central American cichlids. In the shallow waters of their habitat, species of this subgenus have less to fear from finned predators than from winged stalkers. The spiny aspect of these otherwise suitable mouthfuls protects them from being eaten by birds. A similar adaptation is found in other cichlid genera as well (e.g., *Eretmodus* from Lake Tanganyika, *Steatocranus* from Zaire, and *Apistogrammoides* from Peru).

The Atlantic sibling species of *C. nigrofasciatum* is *C. spilurus*. The distribution of this cichlid did not proceed to a similar extent as it did for *C. nigrofasciatum*. The distribution of *C. spilurus* extends from the Rio Motagua in Guatemala to the Rio Prinzapolka in Nicaragua. The presence of *Herotilapia multispinosa* may restrict *C. spilurus* to rocky areas with clear water. The large

Cichlasoma nigrofasciatum female with her free-swimming brood. Photo by Hans Joachim Richter.

A dwarf cichlid from South America, *Apistogrammoides pucallpaensis*. This male is tending his spawn. Photo by Hans Joachim Richter.

ranges of several members of *Archocentrus* allow for geographic variation. Many of these variants are very beautiful and are popular among hobbyists.

Like the insectivorous *Theraps*, *Archocentrus* belongs to the cave-spawners. Pairs may even occupy a territory outside the breeding season. Little is known about *C. spinosissimus*, which has a close relationship to *C. centrarchus*. *C. spinosissimus* is found in Guatemala and was described from the Rio Polochic. *C. spilurus* was encountered in the same system. This led to the wrong conclusion that the Rio Polochic population of *C. spilurus* was synonymous with *C. spinosissimus*. The latter, however, can clearly be differentiated by its high number of spines in the anal fin: 11 or 12. *C. spilurus* has at most nine spines in the anal fin. Of more interest is the fact that *C. spinosissimus* has close affinities with *C. centrarchus*, another species with a high number of spines in the fins. This similarity may favor the idea that *C. spinosissimus* prefers, like *C. centrarchus*, soft bottoms and a weedy environment.

When *Archocentrus* split from *Parapetenia* (= *Nandopsis*) it possibly developed along two different lines: one group formed the cave-breeding *C. nigrofasciatum* branch and the other the soft-bottom frequenting *C. centrarchus* branch. *C. spinosissimus* may even be the sibling species or predecessor of *C. centrarchus*, although presently known distribution patterns refute this idea. The fact that *C. centrarchus* is distributed throughout Lake Nicaragua and its effluent the Rio San Juan points to a probable origin of this species in Lake Nicaragua. Its predecessor, which should have come from Nicaraguan rivers, was successfully conquered at a later stage, when the sibling pair *centrarchus/multispinosa* had evolved.

Another explanation for the missing link between *C. spinosissimus* and *C. centrarchus* could be the fact that the fishes of Honduras and Nicaragua are little known and this area may still "hide" some new cichlid species.

The colored areas in the map indicate the ranges of the two species.

Archocentrus, not growing beyond 12 cm (5 in) length. Their habitat consists of shallow water with a hard bottom made of rocks or pebbles.

Aequidens coeruleopunctatus migrated from Colombia through Panama to Costa Rica. Although the genus _Aequidens_ might have a longer history, _A. coeruleopunctatus_ is certainly not better adapted than the _Cichlasoma_ species from Central America. The advantage of this species over the latter lies in its tolerance of soft and acidic water. The freshwater system and its fauna from Panama and part of Costa Rica have a South American character. This implies soft water, with a pH below 7.

On many occasions Lake Nicaragua acted as a "breeding station" for new cichlids. If these new species proved to be successful they dispersed over a wider area. A transient inhabitant of the lake might have been _C. spilurus_. It is not certain whether the _Archocentrus_ found in the lake, beside _C. nigrofasciatum_ and _C. centrarchus_, belongs to the species _C. spilurus_ or _C. septemfasciatum_. It is, however, clear that _C. septemfasciatum_ did evolve from _C. spilurus_. _C. septemfasciatum_ is found in the lake's effluent Rio San Juan. The distribution of this species encompasses the Atlantic drainage in Costa Rica and Panama. Moreover, a population has been discovered in the Rio Tempisque on the Pacific side of northern Costa Rica. The lower half of the Pacific side of Costa Rica harbors the sibling species of _C. septemfasciatum_: _C. sajica_. Both species are specialized toward a more herbivorous diet. Hence, in the Rio San Juan _C. septemfasciatum_ has more

competition from _Neetroplus nematopus_ than from _C. nigrofasciatum_. The two herbivorous _Archocentrus_ found a way to survive alongside the southern _Aequidens coeruleopunctatus_, which is a general invertebrate-feeder found over soft bottoms. Typically, these two are the smallest

Aequidens coeruleopunctatus migrated from Colombia, through Panama to Costa Rica. Photo by Uwe Werner.

Central American cichlids have a life-long history in waters with a high mineral content and are adapted to such circumstances. The creation of Panama and Costa Rica about three million years ago opened the doors for northern as well as for southern cichlid populations. Because of the soft, acidic nature of their fresh water, Panama and Costa Rica are better suited for South American than Mesoamerican (another way of saying Central American)cichlids.

A. coeruleopunctatus is an open substrate spawner and, as such, is more primitive than Archocentrus. The rivers and streams of Panama containing mineral-rich water harbor cichlids from the Mesoamerican lineage. The soft and acidic waters are inhabited by A. coeruleopunctatus and Geophagus crassilabris. The very advanced mouthbrooding technique employed by G. crassilabris did not favor its dispersal over the continent further north than Panama. The

primitive, in respect to breeding mechanics, Cichlasoma of the subgenus Amphilophus held their grounds in Costa Rica and further north.

Large invertebrate-pickers

Probably before Lake Nicaragua was created, C. friedrichsthalii or its ancestor gave rise to an insectivorous cichlid. The size of this cichlid was the same as its ancestor. To be able to sustain a large insectivore, suitable food must be available in abundance. This implies that other insect-pickers were sparingly present (if at all) in the waters preferred by these descendants of C. friedrichsthalii.

New lakes are an optimal habitat for large cichlids, as are brackish waters. The latter habitat is avoided by most Central American cichlids, but C. citrinellum is still encountered in coastal lagoons on the western side of Costa Rica. Before Lake Nicaragua formed, the ancestor of C. citrinellum evolved from a Parapetenia, possibly from C.

woodringi. Because of the adaptive speciation of these insectivores in the newly formed lakes of Nicaragua, they do not resemble their ancestors. The species that might resemble the first insectivorous descendant from Parapetenia is C. lyonsi. This species is still present in coastal rivers on the Pacific side of Costa Rica. It was separated from the later C. citrinellum by the rising of the Guanacastan Mountains in northwestern Costa Rica. C. lyonsi or its ancestor might once have had a much wider distribution along the Pacific side of Nicaragua and Costa Rica, and it may even have given rise to the insect-picker C. macracanthum, which is encountered in the Pacific drainages of Guatemala. C. lyonsi remained the most primitive of this group and may have been replaced by C. citrinellum in its previous Nicaraguan habitat.

The ancestor of C. citrinellum profited from the creation of the Nicaraguan lakes and adapted

The colored area on the map indicates the range of *Aequidens coeruleopunctatus*.

Aequidens coeruleopunctatus

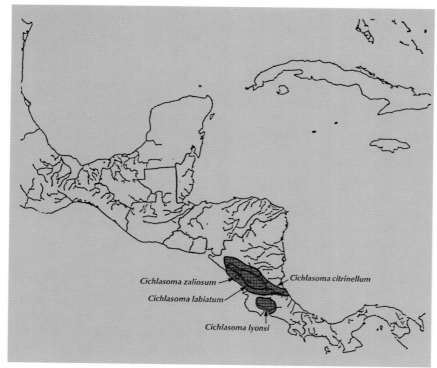

The colored areas on this map indicate the ranges of the 4 species. *Cichlasoma labiatum* and *C. zaliosum* seem to overlap substantially.

The creation of Lake Nicaragua, Lake Managua, and several other lakes took several steps. Initially the whole valley west from the central mountain ridge of Nicaragua was flooded and could have appeared as an extended inland bay filled with sea water. The rising of the continent narrowed down the bay at the east side and turned the water basin in the west into a freshwater lake. This large lake was undoubtedly populated by *C. citrinellum*. Volcanism in the lake itself created isolated small lakes like Lake Apoyo. The Masayan volcanoes separated the present Lake Managua from the larger Lake Nicaragua. A connection between the two lakes still exists, but it is not a wide river as is indicated on many maps. The

Cichlasoma motaguense. Photographed by Dr. Herbert R. Axelrod at the Steinhart Aquarium, San Francisco, when they had a display of many Central American cichlids.

itself to the rocky environment. As a reaction to the impact the lips endured during feeding from the hard, rough substrate, the species developed fleshy lips. *C. citrinellum* became a very successful cichlid and is, in weight, the number one cichlid from Lake Nicaragua. The descent from a *C. friedrichsthalii*-like cichlid is reflected in the ability of *C. citrinellum* (and its close relative *C. labiatum*) to produce white or orange individuals. This character is also known in *C. friedrichsthalii*, *C. motaguense*, and *Petenia splendida*. It seems to be an adaptation to allow the fish to live and breed in deeper regions of a lake. As said before, *C. citrinellum* is one of the first descendants from *Parapetenia* and has, as such, the widest distribution of the large insectivores.

area between the lakes is swampy and may flood only at exceptional occasions of enormous downpours in the rainy season. Migration between the lakes still seems to occur, but the direction is always from north to south. Hence some species may be encountered in Lake Nicaragua that are absent in Lake Managua (e.g., *C. maculicauda* and the shark *Carcharinus leucas*).

The oscillating water level may have played an important role in the creation of two other species of this group, *C. labiatum* and *C. zaliosum*. *C. zaliosum* is found only in Lake Apoyo and has developed from a geographically isolated population of *C. citrinellum*. During its initial isolation *C. zaliosum* underwent

some minor anatomical changes and discovered another food source. It found out that insects that had come to the surface to deposit eggs or had just fallen onto it taste good enough to cram its stomach. On a later occasion, maybe tens of thousands of years later, volcanism in the bigger lakes caused the water level to rise. The high water connected Lake Nicaragua with Lake Apoyo again and a new batch of *C. citrinellum* was introduced. Evolution had taken its course in both populations, and they no longer recognized each other as conspecifics. However, the two species were dining from the same dish, and since they did not accept the other as conspecific, they had to compete with it. Usually this kind of struggle is

won by the most advanced species, but when enough individuals of the less advanced are available, the chances to finding another way out become higher. *C. zaliosum* had learned to feed from the surface and had developed into an open-water species. As such it could avoid contact with *C. citrinellum* from the rocky habitat. Breeding, however, proceeds inshore, but danger is escaped by a quick dash into the open rather than into the rocky habitat.

C. zaliosum is on the very edge of becoming a new species, as natural "hybrids" with *C. citrinellum* are sometimes observed. This reflects its close relationship with the latter. The notion that *C. zaliosum* feeds on zooplankton can be only partially

This is a hybrid of several Central American cichlids. It has the most round body of any of them.

true. Zooplankton is made up of small crustaceans that do not contain enough fat to fill the needs of the fish during breeding, when no food is taken. The striking contrast in the fatty aspect of these two species' flesh reflects the difference in their diets. *C. citrinellum* dines on a wide spectrum of invertebrates from the rocky habitat, the majority being crustaceans, whereas *C. zaliosum* feeds on insects and their aquatic larvae.

Another species that developed in a similar manner but with much less pronounced differentiation is *C. labiatum*. The name indicates an obvious character of this cichlid: fleshy lips. Less well known is the fact that *C. citrinellum* is able to develop similar lips, which may make correct identification difficult. Undoubtedly, the ancestor of *C. labiatum* had time to differentiate into a specialized insectivore by isolation of a population of *C. citrinellum*. When the isolated population was again united with the main population of *C. citrinellum* there were few anatomical differences to set apart the newly developed *C. labiatum* from the re-introduced *C. citrinellum*. The fleshy lips that are always present in specimens from natural populations of *C. labiatum* can be found in some *C. citrinellum*.

Fleshy lips are seen in other species as well, including *C. vombergae*, a *Parapetenia* from the Dominican Republic, and are definitely an adaptation toward the environment. The ability to grow such lips probably is inherited, but the actual growth is induced by the impact they suffer when the fish browses from the rough substrate. Proof of this lies in the fact that any cichlid species that normally develops fleshy lips in the wild will confine the growth of its lips to a mere swollen state

Cichlasoma zonatum. **Photo by Werner & Stawikowski.**

when kept and raised in captivity. This fact seems to disprove the possibility that the lips act as sensory organs in order to "feel" the prey. A sensory organ requires a complex organization in the anatomy of the lip and has to be inheritable. Inheritance is not affected by breeding or by prolonged maintenance of a wild-caught cichlid in captivity. The individuals raised in captivity should thus show lip anatomy similar to wild specimens. Since this is not the case, fleshy lips must be induced by the environment. They are an acquired property of the species; only the ability to grow them is inherited.

C. citrinellum and *C. labiatum* locate their prey by eye before they pick it up. The substrate therefore is deliberately screened for possible food. The substrate can be a smooth layer of biocover or coarse lava. The small holes and cracks in lava stones provide optimum shelter for a wide array of invertebrates. To be able to secure the located prey out of the crack, fleshy lips could be of great help. The lips envelop the entrance of the hole. Expansion of the buccal cavity produces a lowered pressure that forces water to flow into the mouth. This presses the lips tight to the substrate. The inflowing water is thus directed out of the hole and the resident prey is sucked along. In order to prevent the prey from retreating deeper inside the spongy stone, every attack proceeds in a flashing stroke. The fierce impact with the rough substrate further induces the growth of the strained lips. A rougher rock induces a more pronounced lip that can seal off the cracks better.

The presence of spongy rocks of laval origin allows *C. labiatum* to exist beside *C. citrinellum*. Whenever both species occur sympatrically *C. citrinellum* is hindered from developing the fleshy-lip trait, since this adaptation belongs to *C. labiatum*. When the latter is not present *C. citrinellum* may form populations with fleshy lips. The only way to differentiate between a fleshy-lipped *citrinellum* and a *C. labiatum* is to watch them from the top. Seen from above, *C. labiatum* has a more pointed snout whereas *C. citrinellum* has a more blunt snout. Juveniles, however, cannot be differentiated. In my opinion *C. labiatum* is a sibling species that was once distinct but is slowly being integrated into the populations of its predecessor, *C. citrinellum*.

The colored areas on the map indicate the ranges of the four species.

Invertebrate-pickers from the South

From the ancestor of *C. citrinellum*, a cichlid that could have been similar in anatomical features to *C. macracanthum*, a branch was split off that dispersed over the new territory of Costa Rica and Panama. Due to another trait, derived from *C. longimanus*, this line is commonly encountered in clear, running water and over hard bottoms. The most common species of this group is *C. alfari*. Its sibling species is *C. diquis*. The latter is found in a similar biotope on the Pacific side of Costa Rica and western Panama, whereas *C. alfari* is distributed along the Atlantic drainage of southeastern Nicaragua, Costa Rica, and the northwestern part of Panama. It was found that these two species specialized mainly on aquatic insects. *C. alfari* is to be found in the same area as *C. citrinellum* (northern Costa Rica) but their habitats hardly overlap. *C. citrinellum* is more restricted to coastal lagoons and tolerates salty water. In contrast, *C. alfari* prefers mountain streams with cool, clear water.

C. rhytisma is remarkable because of the fact that this species, a direct descendant from *C. alfari*, is found sympatrically with its closest relative. If it were not for the co-existence of both species in one stream, we would have regarded *C. rhytisma* as a geographical variant of *C. alfari*. *C. rhytisma* is stockier than *C. alfari* and has a similar color pattern. In the stream where they are sympatric, the population of *C. alfari* has an aberrant color pattern: the males are adorned with numerous randomly spread black spots on the upper half of the body and rows of spots on the abdomen. It is very unlikely that, without geographical isolation, a new species can be created in the presence of its ancestor. Thus it is virtually impossible that a few individuals of the *C. alfari* population in the Rio Sixaola adapted themselves to a certain niche while the "normal" *C. alfari* was still present. Interbreeding with these supposed variants would obscure any distinctive traits in a few generations and would likely have changed the whole population rather than just a few individuals. If adaptations

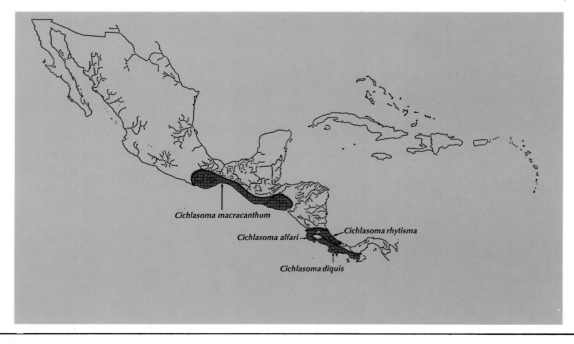

Cichlasoma macracanthum

Cichlasoma alfari

Cichlasoma rhytisma

Cichlasoma diquis

are called for they will occur via evolution. In this process all individuals of the population are involved. An effective branching off of a new species requires selective mating. Only when the variant has developed enough specific characteristics that are not regarded as conspecific by its ancestor has it a chance of becoming a new species. If one of these characteristics includes a new feeding specialization that is seldom employed by its ancestor, a new species is born.

The specific status of *C. rhytisma* is indicated by the fact that no hybridization occurs between the two species (as far as is known). The reproductive isolation evolved during a long period of separate evolution, then volcanism and the rising of the Talamanca Mountains brought both populations together. In numerous similar situations both varieties merge into one species. Here evolution has made two distinct species. We might even imagine that an outburst of a volcano severed a population of *C. diquis* on the other side of the

Above: *Cichlasoma labiatum.* Photographed by Mervin F. Roberts. Below: A pair of *Cichlasoma alfari* spawning. Photo by Uwe Werner.

continental divide. This variant of *C. diquis* would stand a better chance in surviving beside *C. alfari*.

The last species in this group has an uncertain position among these invertebrate-pickers. *C. calobrense* has a robust appearance when compared with *C. alfari*, its overall shape reminding us a little of *C. citrinellum*. *C. calobrense* is found in the same waters as is *Geophagus crassilabris* and may have directed its attention to other prey, like small fishes. As a closer descendant to *C. citrinellum* it could find an existence beside *C. altifrons*, a sand-sifter from the *longimanus* group.

Substrate-sifting insectivores

While the species grouped around *C. alfari* prefer clear, running water with a bottom of wave-washed pebbles, the next group is usually found above detrital substrate. All species of the following group sift the substrate and secure any edible matter from it. The most primitive member of this group is *C. longimanus*. As could be

Cichlasoma facetum hangs its hatchlings on the fine algae. This charming photo by Rainer Stawikowski.

anticipated, this cichlid is distributed over the least competitive region: Nicaragua and Honduras. The occurrence of *C. longimanus* on the Pacific as well as on the Atlantic side of the subcontinent is remarkable.

C. longimanus is a general bottom-feeder but prefers soft bottoms in warm, slowly moving water. Its food consists of anything edible found in the sediment and ranges from snails and crustaceans to diatoms and algae. As such it could spread over large areas since

competition for food was not present. On the Pacific side of its distribution competition is still not evident, at least not from a known cichlid species. At the northern edge of its territory *C. macracanthum* could overlap part of *C. longimanus*'s diet.

In Nicaragua, *C. longimanus* is confronted with more cichlids at the dining table, most of them derivatives from the "species factory," Lake Nicaragua. As a consequence the Pacific *C. longimanus* grows to about 18 cm (7 in) in length, while the Atlantic

populations succumb to the competition and produce individuals with a maximum size of around 12 cm (5 in).

Lake Nicaragua yielded a new species in this group: *C. rostratum*. This cichlid is specialized for the sandy substrates, whereas its ancestor, *C. longimanus*, was pushed into the muddy areas. *C. rostratum* may grow to 30 cm (12 in) in length and belongs among the most beautiful cichlids, including not only those from Central America. Like its namesake from Lake Malawi, the male rostratum attires itself in a striking blue-black hue speckled with colorful spots. Due to the coarse sand chewed by this cichlid, its lips are swollen. Such reinforced lips withstand the straining that arises from the plunge into the sand better than do conventional lips. Again, individuals under natural conditions show thicker lips than do most tank-raised specimens. If we provide this species ample fine sand it may develop the same pronounced lips as in the wild. *C. rostratum* is found mainly in the large lakes of Nicaragua, its place of origin. Some individuals may stray and descend the Rio San Juan.

Migrating *Amphilophus* have populated Costa Rica and formed the basis of a new species: *C. altifrons*.

Some populations of the wide-spread *C. altifrons* have a striking resemblance to *Geophagus brasiliensis*. The fact that the latter is to be found thousands of miles away excludes a possible relationship. Still, it is remarkable how cichlids may solve a feeding problem and come up with similar species. It seems that there is wide variation in cichlids but that the different types of traits are well-defined.

C. altifrons is dispersed over

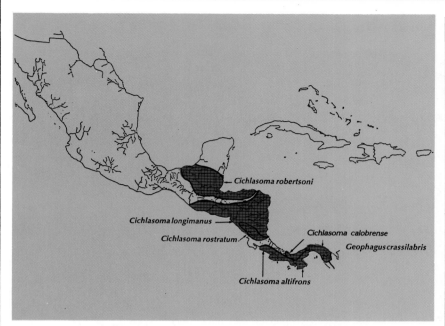

The colored areas on the map indicate the ranges of the six species.

the Pacific side of Costa Rica, Panama, and even into Colombia. Some scientists believe that *C. calobrense* is the sibling species of *C. altifrons*. However, they are both encountered in the same river systems! Both are found in the Rio Tuira and Rio Bayano of the Pacific drainage in Panama.

Substrate-sifting species from the North

C. longimanus also gave rise to a new and more advanced species in the north, *C. robertsoni*. Up to now these two species have not been observed in the same stream. The borders of their ranges run through central Honduras. *C. robertsoni* is encountered only in Atlantic drainages. Probably *C. longimanus* had a distribution encompassing the complete Nuclear Central America and created *C. robertsoni* at the other side of Bahia de Amatique. The displacement of its ancestor in western Honduras could indicate the success of this new species.

The snout of *C. robertsoni* is very elongated, with a terminal mouth. Werner and Stawikowski observed this species in its natural habitat and found that it preferred a soft bottom with a thick layer of decaying leaves and debris. The lips of this species are not thickened, which corresponds to the soft substrate

Cichlasoma robertsoni. Photo by Conkel.

from which it feeds. The eyes are placed at a high position in the head, which allows the fish to dip its mouth quite far into the sediment. The substrate is chewed and the edible material is retained. *C. robertsoni* is rather variable over its wide distribution. Probably it was not challenged by a similar species in the new area of Yucatan/Guatemala.

C. margaritiferum is known only from the type specimen of unknown origin. Its anatomy resembles that of *C. robertsoni* and the species might well be a geographical variant of the latter. Until new material is found this taxon may be grouped with *C. robertsoni*.

C. robertsoni undoubtedly gave rise to a whole new group of cichlids in the biogeographical region of Yucatan/Guatemala, namely the subgenus *Thorichthys*.

Substrate-filterers from shallow water

In Yucatan/Guatemala we have met insectivores from the subgenus *Theraps*. Their average length is about 20 cm (8 in), and thus they are not to be found in the extreme shallows. Parallel to the evolution of *Archocentrus* in Honduras/Nicaragua, *Thorichthys*

The most spectacular of all Central American cichlids might well be the firemouth cichlid, *Cichlasoma meeki*. It has been bred in Florida for almost 75 years and is known for its hardiness and pugnacity. In this remarkable photograph, Hans Joachim Richter shows a male guarding its free-swimming fry. When threatened it puts on this beautiful display of red.

adapted itself to food-rich sections of the streams. In contrast to the cave-breeding *Theraps* is *Thorichthys*, a typical open substrate spawner. *Thorichthys* is, however, a common sight in its habitat, which includes the shallow shores of all types of rivers. Its depth distribution ranges from 10 to 150 cm.

Direct descendants of *C. robertsoni* could be *C. affine* and *C. meeki*. Both are considerably smaller and may reach a maximum length of 15 cm (6 in), whereas *C. robertsoni* may grow to 25 cm (10 in). Moreover, *C. meeki* and probably other *Thorichthys* may breed at a length of just 5 cm (2 in)! This means that these cichlids can populate extremely shallow water away from most predators. Feeding behavior of *Thorichthys* may vary not only among species of the group but also in respect to *C. robertsoni*. The latter plunges its terminal mouth deep into the soft bottom while it keeps its body in a slanting position. *Thorichthys* has a less symmetrically pointed snout and eats at a more horizontal position. This allows foraging in body-deep water. *C. robertsoni* has been observed together with *C. meeki* or *C. helleri* but preferred the slightly deeper regions.

C. meeki, the common firemouth cichlid, is encountered in a relatively large area and has even adapted to brackish lagoons in Belize. Its diet consists of invertebrates and vegetable matter recovered from the soft substrate. *C. affine* has a similar feeding pattern. Up to now, *C. meeki* and *C. affine* have not been found together.

The long snout of *C. ellioti* from some populations reminds us of its ancestor, *C. robertsoni*. Also, *C. ellioti* is involved in feeding

Above: Firemouth cichlids, *Cichlasoma meeki,* normally carry the red throat and belly but are much more intensely colored when breeding. Below: The colored areas of the map indicate the ranges of the two species.

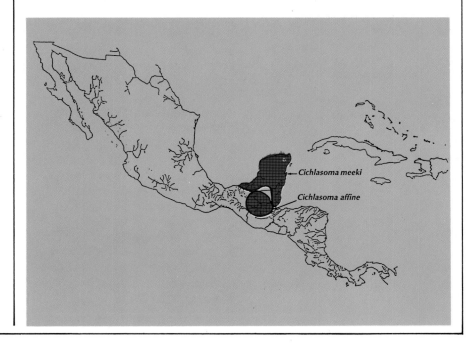

Cichlasoma meeki

Cichlasoma affine

Cichlasoma meeki has probably not been imported from its natural range for many years because it is so easily bred from previously bred generations. It is amazing that these firemouth cichlids are still popular since they are very pugnacious. The extendible snout, as shown by the lightly colored male in the foreground, indicates an ability to suck in small fishes.

from the soft bottom. Other populations of this species are specialized for picking invertebrates from the biocover in shallow rocky areas. Their snouts are much less elongated. The screening of the biocover on the rocky substrate is performed by other species in this group as well. *C. socolofi, C. callolepis,* and *C. helleri* normally are found in clear water over small rocks and stones. The shallowness of the water allows for a rich algal growth. The algal blanket provides optimum shelter for aquatic insects and crustaceans. Since this nutritious prey is left untouched by other cichlids, the insect-picking *Thorichthys* (*C. helleri, C. callolepis, C. socolofi,*

C. aureum, and *C. ellioti*) thrive on it. These five species have a disjunct distribution and are typically encountered in clear mountain streams with rock-bestrewn bottoms. Although these habitats provide ample shelter, progeny are cared for in the open water.

C. aureum is the only species of the group of five that specializes in feeding on snails. It has enlarged, molar-shaped pharyngeal teeth. Undoubtedly, the presence of the insectivorous *C. spilurus* in the same area has forced *C. aureum* to restrict itself to snails.

Another feeding specialist is *C. pasionis,* which was named after the stream where the type

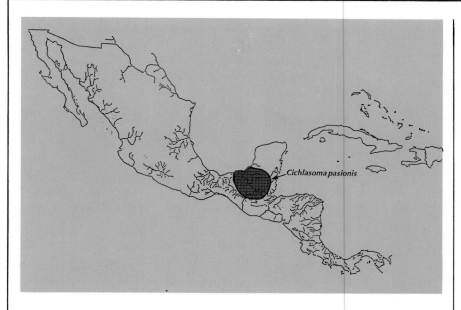

specimen was captured. Its distribution covers a larger region, and the species has been observed with *C. meeki* and *C. affine*. It seems to have a similar range as *C. helleri*, sometimes resulting in the occurrence of three species of this group being found together. *C. pasionis* feeds predominantly on hard-shelled insects that are recovered from sandy patches in the shallow waters. Small fishes will occasionally be taken, too.

Thorichthys is a relatively new subgenus covering four different feeding patterns. The group is focused on shallow areas where competition mainly is with other species of the same group. The most primitive members of the subgenus (*C. affine*, *C. meeki*, and perhaps *C. ellioti*) screen the soft bottom for something edible, whereas *C. pasionis* prefers sandy inserts in the rocky areas. The most advanced group (*C. socolofi*, *C. callolepis*, *C. helleri*, and some populations of *C. ellioti*) are observed to pick their prey selectively from the biocover. Nevertheless, they are occasionally seen digging in the bottom. At the eastern edge of its range *C. ellioti* might find itself in

The map above indicates the range of *Cichlasoma pasionis* in the colored area.

The map below shows the distribution of five species of Central American cichlids.

a rather competition-free area and may occupy several habitats. As a consequence, many different variants can be observed. Some of them have elongated snouts and are encountered above soft bottoms, while others have relatively short snouts and slender bodies and are observed in rocky regions.

Snail-feeding might have arisen from the heavy competition present in the area where *C. aureum* is found. The adaptation to crush snails is not a big step if starting from an insectivorous feeding pattern. The enlargement of the pharyngeal teeth usually coincides with an enlarged set of chewing muscles. It has been shown that the active crushing of snails induces the growth of these muscles. If a predominantly snail-feeding cichlid is not allowed molluscan food, the muscles of its pharyngeal apparatus will partially deteriorate.

Cichlasoma from South America

Another area where cichlids suffered little competition is the northwestern corner of South America. After the island bridge

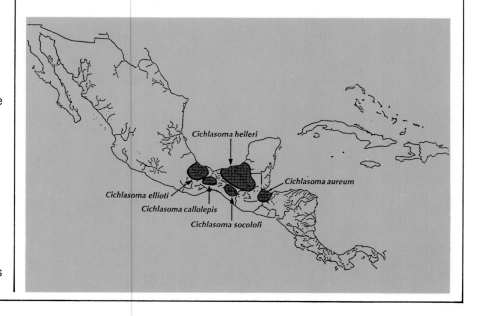

between Nuclear Central America and South America shifted eastward to form the Antilles, some cichlids remained at the northern coastline of South America. These cichlids were isolated from the rest of South America by the Andean mountains. Primitive *Cichlasoma* were probably also present at the other side of the Andes, but over time they were displaced by species from the Amazon basin.

Three species developed in this isolated northwest corner: *C. atromaculatum*, *C. ornatum*, and *C. festae*. The last two resemble each other, which corroborates their slow evolution. *C. atromaculatum* had to cope with an invasion of Central American cichlids when Panama and Costa Rica connected the two continents. At both margins of their distribution *Cichlasoma* suffers the invading *Aequidens*,

PANAMA

Cichlasoma umbriferum

Cichlasoma atromaculatum ⟶

Cichlasoma ornatum ⟶

Cichlasoma festae ⟶ ECUADOR

PERU

The fish above is a mouthbrooding *Geophagus steindachneri* that also develops a nuchal lump as this male shows. Photo by Hans Joachim Richter. Left: This map shows the distribution of some South American *Cichlasoma*.

which are better adapted to soft and acidic water. Food usually is not as abundant as in the Central American rivers, so an aggressive trait had to be developed to withstand the competition. *C. festae* shares its habitat with *Aequidens rivulatus*, the green terror. Both species were originally open substrate breeders, but the fierce competition for spawning sites forced *C. festae*, a member of *Parapetenia*, to breed in caves.

When the two Americas were again united, *Cichlasoma* from South America could also enter Panama and extend farther north. However, there they would have met better adapted members of the genus, leaving little space for the more primitive species to live. One species could have originated in South America: *C. umbriferum*. Like *Petenia splendida*, a *C. atromaculatum*-like cichlid could have specialized in stalking characoid fishes. This could have paved its road into Panama. *C. umbriferum* has a protrusible mouth similar to *Petenia* and can grow to an unbelievable size of 80 cm (30 in) (reached in captivity); it is not clear whether this species could attain these dimensions under natural conditions.

The fish above is an African species, *Cyphotilapia frontosa*, that has a nicely developed lump on its head. Photo by Dr. Herbert R. Axelrod at the Berlin Aquarium in Germany.

The fish below is an *Aequidens "rivulatus"* showing a nuchal protrusion. Photo by Werner & Stawikowski.

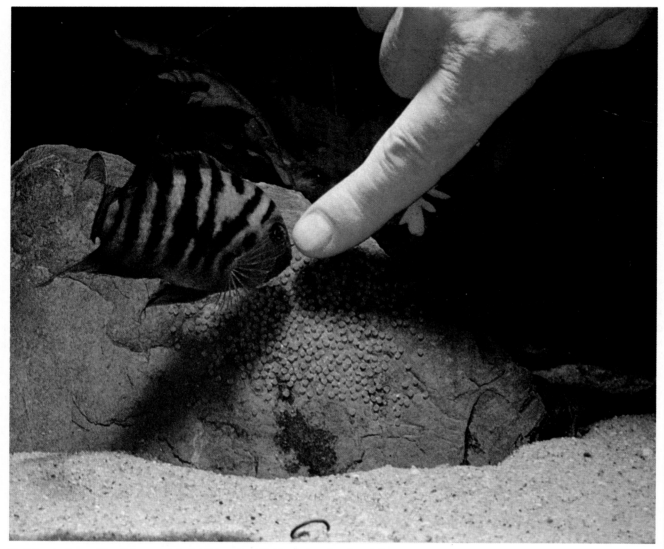

Nothing is feared by this convict cichlid male, *Cichlasoma nigrofasciatum,* as he protects his spawn. The teeth of this fish are inconsequential to your finger, so show off to your friends about how protective your fishy pets may be. This kind of show endears aquarium fish to the hearts of nature lovers. Photo by Ruda Zukal.

At the beginning of cichlid evolution, the fishes were large and enough food was available to maintain them that way. It took several years before a cichlid reached maturity and added its breeding efforts to the perpetuation of the species. Generations passed at a slow pace and the competition was too small to stress the population. After geographical isolation and rejoining of several species, all efforts were directed toward a better adapted fish that could withstand increasing competition. The high degree of adaptability of the cichlids led to denser populations with less food available. This could have led to fry feeding from the skin of their parents. The hungry fish acquired some aggressive traits in gaining food. The successful species changed their diet and reached maturity at a smaller size. This shortened the generation time and thus could speed up evolution.

The increased number of cichlids in the habitat diminished the number of available spawning sites. A new battle started: finding a suitable place to rear offspring in security. Some smaller species hid their progeny in caves, others in the mouth.

Pair formation

Central American cichlids are solitary fishes with a rather strong territorial behavior. Pair formation, if such ever exists in these species, will take place only during breeding. The more primitive species display sexual dichromatism. This means that the female and male have clearly different colors during breeding. This reflects the different tasks the parents have in rearing offspring. The larger male has to defend his territory in which the female, usually dark in color, was allowed to deposit eggs. The

A magnificent male *Cichlasoma octofasciatum*. Photographed by Dr. Herbert R. Axelrod at the Steinhart Aquarium. Note the small teeth protruding from the fish's lip.

female remains at a much closer distance to the fry.

The species that don't show this dichromatism during breeding (they usually do change color, but not to differentiate male from female) defend the territory in a joint effort. This is of great importance in those areas where competition for spawning sites is high. The joint defense of the territory strengthens the pair-bond. In captivity these species should be bred in the presence of other fishes. These fishes will become the targets of the parents, which would otherwise release their aggression on each other. A small pair of *C. nigrofasciatum* is even successful in chasing large adult *C. dovii* from their premises.

In some habitats spawning sites are difficult to come by. *C. citrinellum* solved this problem by forming a pair before securing a cave to breed. The joint action of the pair results in a stronger bond and a better chance to obtain a suitable site.

Competition can be so great that it is more economical to defend a territory at all times. This is seen in *Neetroplus*

Cichlasoma salvinii male guarding its flock of free-swimming fry. Photo by Rainer Stawikowski.

nematopus, where a kind of colonial organization assures the species of ample breeding space. Not one pair but several pairs may fend off intruders.

Finding a place to breed

The primitive *Cichlasoma* are open-substrate-spawners. The large male screens the habitat for a spawning site, usually a flat horizontal surface, and attracts a female. The female relies on the defending abilities of its current spouse and directs her attention to rearing fry. The territory is small as the residents have little to fear from other inhabitants of the biotope.

When competition from other species increases, so does the size of the territory. This is especially true in clear waters where food is relatively scarce. The future progeny need ample

foraging ground. When the population density increases the fry succumb to heavy predation, even if the parents belong to the largest inhabitants of the biotope. This has caused *C. dovii, C. festae,* and other *Parapetenia* to breed in caves. During evolution only the large *Cichlasoma* were able to protect their fry in open water (e.g., *Parapetenia* and the herbivorous *Theraps*). All other species found spawning sites in caves, shallow water, or developed other techniques.

Small species had more options to protect their offspring. Some of them moved to very shallow water where the larger predators (feeding on the parents) could not venture. On a rocky bottom they would breed in small caves; on a soft bottom they would probably hide their progeny among the weeds.

Some species, such as *C. nicaraguense* and *C. longimanus,* forage from the sandy or muddy bottom but need shelter to breed. This poses a great problem in those areas where rocky habitats are heavily populated. For *C. nicaraguense* it means a fierce fight with the successful *N. nematopus.* These fights are usually lost by *C. nicaraguense.* This could result in the remarkable fact, observed by McKaye, that the males defend fry of another species (*C. dovii*). *C. nicaraguense* is not a real

---

Cichlasoma nigrofasciatum spawning on the face of a rock in the wide open spaces even though more secretive places were available. These fish have a strong protective instinct. Photo by Ruda Zukal.

cave-breeder and is sometimes observed spawning in pits between rocks. However, it behaves like a true cave-breeder. The female selects the spawning site and attracts the male. This is in strong contrast with the open substrate spawners where males select breeding sites. This may indicate that *C. nicaraguense* once was a true cave-breeder but that increased competition pushed it to deeper levels.

A part of the *C. citrinellum* population in the lakes of Nicaragua breeds at deeper levels. At these levels daylight hardly penetrates. As an adaptation, about 10% of the total population of this species has an orange coloration. The greater part of these aberrantly colored individuals spawn at deeper levels than the gray-beige fish. If selection were to become stronger, probably all individuals would breed at deeper levels and the whole population would eventually change into orange fish. This would not mean that a

In the upper photo, *Cichlasoma nigrofasciatum* spawns in a more secretive fashion inside an overturned flowerpot. This female is guarding her eggs. The fish is called the "convict cichlid" because of its black stripes. Photo by Arend van den Nieuwenhuizen. In the photo to the right, a pair of pink convicts, *Cichlasoma nigrofasciatum* also chose to spawn in a flowerpot. Photo by Ruda Zukal.

new species was born, but simply that this population of *C. citrinellum* turned into orange fish.

Concealing eggs

Eggs of substrate-breeders are permanently stuck to the rocks or plants, thus they are vulnerable to predation. The protection of eggs starts with the selection of the breeding site. Eggs of open-substrate-breeders are transparent or sand-colored. Even in a tank at close range, these eggs are difficult to discern. Some species, like *C. longimanus* and *C. altifrons*, cover the eggs with sand to hide them completely from the predator's eye.

Eggs of cave-breeders are not transparent because in the dim light even the female would not be able to see them. The whitish or yellowish eggs are usually stuck to the ceiling of the cave because nocturnal predators such as catfishes are not likely to screen this part of the cave.

Some species affix the eggs in a vertical crack or slit in the rocks, like *C. citrinellum*, or heap the non-adhesive eggs in a small pit, like *C. nicaraguense*.

The best way for cichlids to protect eggs from predation is to brood them in the mouth. The two great advantages are movability and optimal concealment. This breeding behavior could only have developed in areas where severe breeding site competition existed. Bare and sandy river beds are a likely site of origin of mouthbrooding. Only one mouthbrooding cichlid is known from Central America, *Geophagus crassilabris*, and this species originated in South America. If there is any Central American cichlid on the verge of becoming a mouthbrooder it will be *C. nicaraguense*. This species forages on the lake floor but still breeds in the rocky area. Its non-adhesive eggs that can be taken up and moved to another, safer place are remarkable.

Eggs and hatching

Before a pair is able to spawn, both male and female have to go through hormonal development in which the eggs ripen in the ovary and the male produces fertile sperm. Only when both fish have arrived at the same hormonal level can a successful brood be produced.

The eggs of all substrate-breeders are covered by a sheath of sticky sugar compounds that harden a few minutes after oviposition. Very thin threads extending from the sheath affix the just-laid egg to the substrate.

Geophagus are a genus of mouthbrooding cichlids from South America with only one representative in Central America (not this one shown here). They store the eggs and harbor the fry in their mouths to protect them . . . until the fry no longer fit! Photo by Hans Joachim Richter of *Geophagus balzani*.

Cichlasoma sajica spawning in a cave they dug out from under a rock. Photo by Uwe Werner. Right: A closeup of the eggs and empty egg cases. The fry are stripped from the egg cases by the parents. These eggs happen to be from the African species *Neolamprologus leleupi*. Photo by Hans Joachim Richter.

These threads are made from the same material and, therefore, lose their adhesive properties after a few minutes. When the sheath has hardened it becomes impossible for sperm to penetrate it, thus fertilization has to take place immediately after the eggs are deposited. This is the reason why the female does not lay a large batch of eggs at once, because before the male could fertilize them they would be impenetrable. Small numbers, from one to 20 eggs, are deposited in a single row. The male passes its vent over the just-laid eggs and expels his sperm. In the sheath are several pores through which a sperm-

VARIOUS COLOR VARIATIONS AND COLOR PHASES OF *CICHLASOMA SAJICA*. All photos by Uwe Werner.
Top, facing page: Young male showing minimal coloration.
Center, facing page: Female in full color.
Bottom, facing page: Yellow color variety, female.
Left: A breeding pair with the male in the foreground. Notice that they are protecting their free-swimming offspring and therefore have a distinctive dark color.
Above: The same pair in normal coloration.

Cichlasoma sajica with their free-swimming fry. The colors of the fish change with their mood as well as with their breeding progress. Photo by Uwe Werner.

head can find its way to the egg. Once one sperm has penetrated and fertilized the egg, the pores close and penetration by other sperm is thus prevented. After a short while the sugar compounds (mucopolysaccharides) harden and give the egg an important protection against mechanical stress such as is caused by sand grains or water currents. Of more importance is the impermeability of the sheath to bacteria. Eggs of substrate breeders are immobile and thus are very prone to deterioration. Debris can collect between them and house bacterial infections. The task of the parent(s) is to scrutinize the eggs and remove any dirt that might have collected between them. They do this with fin-fanning. Besides removing the dirt they also pass fresh, oxygen-

rich water along the eggs. Unfertilized eggs usually get a whitish appearance that is certainly not fungus but a coagulation of yolk proteins. Fungused eggs appear after a bacterial infection has made a lesion in the sheath. This might happen in a dirty environment, such as an overcrowded aquarium. If fungus gets hold of some eggs it may occasionally overgrow the whole brood, but normally it will never come to this and the eggs hatch after a few days of development.

Before the larva leaves the egg the shell is opened by an enzymatic process. As soon as the first crack appears in the shell the parents get nervous and prepare themselves for larvae-transfer. During the development of the egg the hard shell

prevented any odor from diffusing through the sheath. This is an important protection against predation. A predator might smell the eggs and dine on them during the night, when the parents were unaware of the predator's presence. Developing eggs don't smell. The unmistakable odor of fry is released through the first crack in the shell and activates the parents. This is a most vulnerable stage in the development of the young. Usually the male assists the female in collecting larvae from the eggshells. The parents "chew" the larvae free from the shells, and in fact they suck the larva out of its opened eggshell. Particularly in open-substrate-breeding cichlids in which the brood is deposited on a horizontal surface do the parents have to assist the larvae to leave the shell. It is not clear if the parents actually break the eggshell, but this is very doubtful. The weak larvae are not able to wriggle out of the eggs themselves, and the parents also are afraid that predators will smell them and energetically transfer them to a previously dug pit. Larvae from cave-breeders are much more easily released from their eggs, as they are affixed on the ceiling or on the side of the cave. These eggs are also less prone to deterioration as dirt will hardly collect among them. Thus they are less frequently fanned by the female.

In some open-substrate-breeding *Cichlasoma* this behavior has consequences when we try to hatch the eggs away from the parents. In order to give the larvae a better chance to escape the eggshells, we have to place in a slanting position, the stone or other object on which the eggs are deposited. Never direct the airstream straight onto the eggs but a few inches from them. This will create a water current that is far less destructive than air bubbles. At the time of hatching you could help the larvae by shaking the stone until all wrigglers are released.

Feeding fry

Fry of most species of *Parapetenia* and *Amphilophus* survive the critical initial stage of their life by feeding from the mucous film on the parents' skin. These primitive cichlids have found an elegant way to escape having to enter the predator-rich region in the upper layers of the water. Only large fish with sufficient reserves are able to breed in areas where food is scarce, even food for fry. These cichlids usually don't eat during breeding and let the fry graze on their flanks when the supply of other food has dropped to zero.

Only the small and successful species from the upper layers of the biotope are able to dine on the available plankton. Some larger species (as of *Amphilophus*) chew up excessively large food morsels and spit the fragmented pieces in front of the fry.

When fry dine on plankton they move in the open water column, thus creating an ideal target for predators. It is difficult for small parents to defend fry in a cloud, so this behavior is not observed among the smaller *Cichlasoma*. A

Cichlasoma cyanoguttatum is very similar to *C. carpinte*. This is a female guarding her free-swimming fry. Photo by Uwe Werner.

few days after the fry become mobile, the territory is too small to feed all of them. Since the small parents are not able to project a larger area, they lead the fry to other foraging grounds. In fact, they move the territory, which is an important step in the breeding evolution of cichlids.

Defense of the Fry

The most primitive group of Central American cichlids in respect to breeding is *Parapetenia* (= *Nandopsis*). With very few exceptions, they all belong to the open-water-breeders. There is a clear distinction between the male and the female. The larger male has to defend the temporarily mutual territory, while the smaller and sometimes differently colored female stays close to her brood. These cichlids spawn on top of a horizontal slab or attach the eggs to driftwood or underwater roots of trees or other terrestrial plants. Usually the spawn lies exposed on all sides and is therefore vulnerable to predation. However, most members of this subgenus are quite large and are able to protect their broods against intruders. These cichlids do not guard their offspring long enough to guarantee them a reasonable chance of survival, however. The large size of the parents calls for a high food intake that is drastically reduced when breeding. In order to survive they have to abandon their progeny at a relatively early stage. As compensation they produce a tremendous number of fry that are able to maintain the species.

Normally the male loses interest first and leaves the female with the fry. At this moment the pair-bond is broken, and it cannot be maintained under artificial conditions. The female continues to defend her offspring, but gradually the fry

Cichlasoma spilurus spawning. Photo by Ruda Zukal.

C. spilurus with an old male dominating the scene. Photo by Ruda Zukal.

Cichlasoma spilurus female guarding her eggs. Photo by Ruda Zukal.

wander astray or are eaten. When all the young have disappeared the female resumes her normal coloration and pattern. The presence of fry induces in the female her strong defensive behavior, but when their numbers decrease so do her ardent attacks against threats. In the confinement of the aquarium the deserting male may regard the female, when she has assumed the normal pattern, as an intruder into his territory and chase her into a corner. This is especially true for those species that suffer high competition for food and are therefore strongly territorial. Notorious for this kind of behavior are *C. dovii* and *C. umbriferum*. A healthy and well-fed female, however, might again start a breeding cycle and elicit a rather peaceful reaction from the male after the previous spawning has been terminated.

In order to compensate for the high losses of fry after they wander from the female, some species may produce more fry. *C. managuense* is known for its huge broods that may number some 5000 fry! Higher quantities of fry will increase the number of predators and will not be the ideal solution of this problem. A much better solution is a decrease in incubation time of the eggs, which is achieved in, for instance, *C. dovii*. Instead of 12 to 14 days being needed for the development of mobile fry as in, e.g., *C. friedrichsthalii* and *C. tetracanthus*, the fry of *C. dovii* and *C. umbriferum* can swim off the bottom in eight days. These eight-day-old fry are the same size as the 14-day-old fry of *C. tetracanthus*. Hence *C. dovii* is a better adapted cichlid. Its large size enables a fruitful defense of a large territory in which the impressive cloud of fry may find enough food to survive the first few days. Nocturnal catfishes

may eventually make all these efforts useless, but enough fry withstand the struggle to adulthood and maintain the species in a competition-rich environment.

Some females (and also males) attain a dark, almost black, coloration when the fry become mobile. This is especially true for species that are regularly found in clear waters or on flat bottoms. The possible reason is that fry, when threatened, dash for a safe place on the bottom. When the water is clear and the bottom consists of pale sand, they undoubtedly would dash for any dark spot. Therefore, the breeding female carries this dark "cave" on her body.

Above right: *Cichlasoma spilurus* assumes various poses in order to lay and fertilize their eggs. Photo by Ruda Zukal. Below: Dr. J. Vierke, one of the world's leading authorities on cichlids and aquarium management, photographed this color phase of *Cichlasoma spilurus* guarding its young. Most cichlids get much darker when they are tending their free-swimming fry.

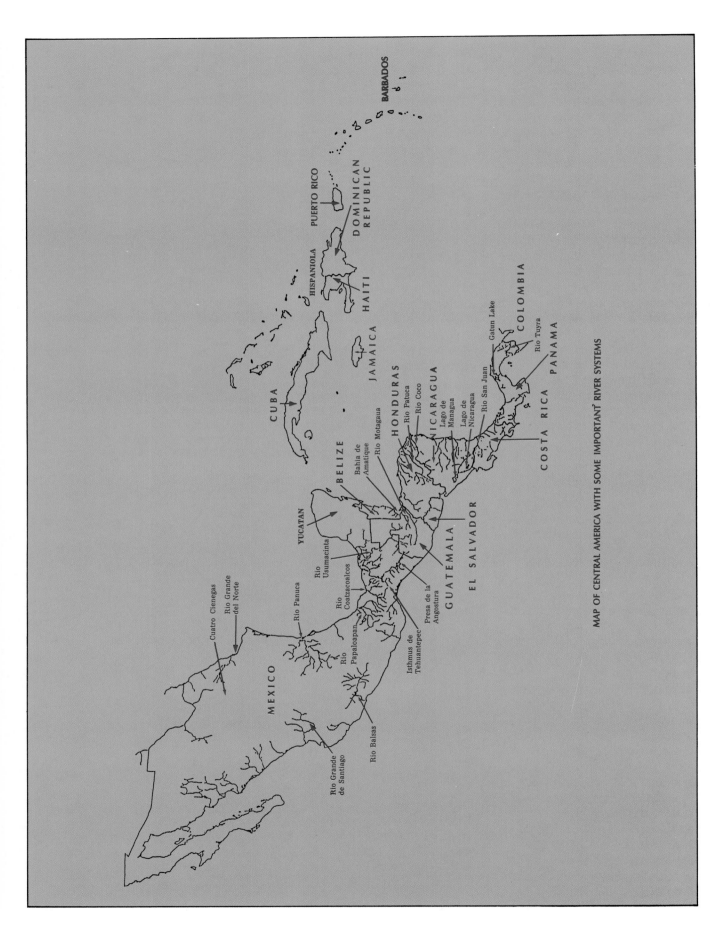

MAP OF CENTRAL AMERICA WITH SOME IMPORTANT RIVER SYSTEMS

88

Cichlasoma (Theraps) synspilus HUBBS, 1935

In the photo above: A male *Cichlasoma synspilus* in breeding condition, about 25 cm (10 inches). It was on display in Nancy Aquarium, France, and was photographed by Dr. Denis Terver. In the lower photo: An immature *Cichlasoma synspilus*. Brewer photo.

Cichlasoma (Theraps) synspilus

C. synspilus has its home in Mexico and is encountered in the drainages of the Rio Usumacinta and the Rio Grijalva. Two color variants are known: the most commonly kept is the red form, but the less conspicuous green/blue form is encountered in most parts of the range. The origin of the red form has not yet been ascertained. The water of its habitat may vary from fresh water with a conductivity of 150 microSiemens to brackish lagoons with a mineral content of 15 grams of salt per liter. Most of the waters harboring C. synspilus are turbid or even murky.

In the photo above: A male, 20 cm (8 inches), *Cichlasoma synspilus*, found at the Nancy Aquarium, France. Photo by Dr. Denis Terver. This is the green variety. Below: Another color variety of *Cichlasoma synspilus*. Photo by H. Ross Brock.

Nevertheless, these cichlids are to be found at high densities and are prized delicacies commonly found in the food markets.

The food of *C. synspilus* consists of vegetable matter. Fruits and half-rotten leaves usually are taken. In captivity this species relishes everything offered, and you have to be prudent with beefheart and earthworms. A steady supply of commercial food pellets is recommended. Beefheart may be added but should not be the main part of the diet. The maximum length of a male is 35 cm (14 in), but 25 to 30 cm (10-12 in) is more common. *C. synspilus* may breed for the first time at a length of about 15 cm (6 in) for the female and 20 cm (8 in) for the male. Under natural conditions this size is attained in about one year. In captivity the juveniles may acquire their beautiful colors in the tenth month.

A magnificent male *Cichlasoma synspilus* with an exaggerated nuchal hump. Photo by Hans Mayland.

An interesting spawning sequence of *Cichlasoma synspilus*. The male, in the photo above, begins the spawning cycle by scrupulously cleaning the selected spawning site. Below: The eggs have been laid and the male fans fresh water over them. Photos in this series by Rainer Stawikowski.

The female *Cichlasoma synspilus* continues laying eggs. They lay hundreds of eggs with some pairs producing as many as 1,000 eggs. Immediately after the female lays some eggs, the male (below) comes over to fertilize them. Usually he waits for the female to leave the spawning site before fertilization takes place.

C. synspilus should be kept in roomy quarters. A tank of 500 liters (130 gallons) will suffice as a breeding aquarium but a community aquarium should measure at least 800 liters (200 gallons) and may be stocked with two or three pairs ranging from about 20 to 25 cm (8-10 in) long. Needless to say, only one pair of *C. synspilus* can be maintained per breeding tank. *C. synspilus* can be kept with other herbivorous *Theraps*, such as *C. maculicauda* and *C. bifasciatum*, but not with *C. melanurum*, its sibling species.

C. synspilus is a substrate-spawner that deposits its eggs on a horizontal slab. To induce breeding you might raise the temperature from 75° to about 83°F (24-29°C). You may use a clear divider between the pair, but the relatively docile nature of this cichlid allows for a more natural breeding method. Although pair-bonds do not exist outside the breeding period, the typical mouth tugging is seldom seen in *C. synspilus* when the pair has been kept together. When a horizontal hard surface is missing in their artificial environment they dig into the sand or gravel until the glass bottom is reached. You can accommodate the breeding pair with a flat stone placed in the center of the pit.

Oviposition normally takes place in the evening. The female

In the Moscow (USSR) Aquarium at the Moscow Zoo many strange hybrids are found because Moscow aquarists do not have access to many fishes. The *Cichlasoma* shown here look like *C. synspilus* but resulted from breeding two related, but not identical, species. Central American cichlids are easily hybridized, but in most cases the hybrids are not interesting, neither in behavior nor in color. Photo by Dr. Sergei Kochetov, Director of the Moscow Aquarium.

A young, but sexually mature, male *Cichlasoma synspilus.* The nuchal hump on his forehead is just beginning to show. This fish is about one year old. Photo by Rainer Stawikowski.

A fully mature female in lovely coloration. Only one pair of *Cichlasoma synspilus* should be kept in a breeding tank at any one time as they are very aggressive when spawning. That's why they are so successful in the survival of the fittest tests of evolution. Photo by Rainer Stawikowski.

A beautiful *Cichlasoma synspilus* with his free-swimming offspring. This is a color variant. Different color varieties exist in a basic red pattern and a basic green pattern. No one seems to know the origin of the red pattern. Photo by Rainer Stawikowski.

may deposit over 1000 eggs that are carefully laid to cover an area of about 15 by 15 cm (6 x 6 in). Now and then the male interrupts the laying female and fertilizes the eggs. During the following days *C. synspilus* becomes somewhat aggressive toward intruders, but the small size of the male's territory allows for a reasonable harmony among inhabitants.

In two days the eggs hatch and are removed to a previously dug pit. During the next days the wrigglers are frequently moved from one shelter to another. One week after hatching the fry are free-swimming, but they are very small in comparison with fry of other *Theraps* species. They should be fed with *Artemia* and will grow fast.

Cichlasoma synspilus. This is a 10-month-old male photographed by Hans Joachim Richter.

Cichlasoma synspilus. Photographed by Conkel.

Cichlasoma synspilus. Photographed by Hans Mayland.

The *Cichlasoma synspilus* shown on this page are all color varieties from various rivers or lakes in Central America, according to the photographers.

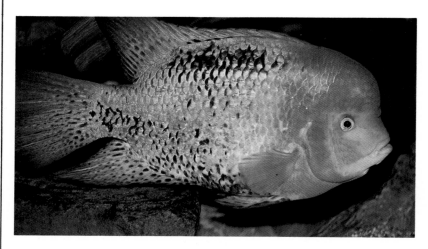

The two fish shown on this page are further color variants of *Cichlasoma synspilus.*

A mature male *Cichlasoma synspilus.* Photo by the author, Ad Konings.

A young male *Cichlasoma synspilus.* Photo by Conkel.

Below: A closeup of the ovipositor of the female *Cichlasoma synspilus* as she deposits eggs. It is amazing that the eggs are laid next to each other and not one atop the other. Photo by Rainer Stawikowski.

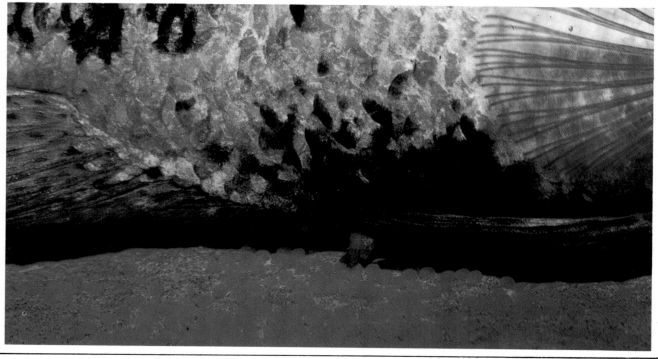

Cichlasoma (Theraps) maculicauda (REGAN, 1905)

An adult male *Cichlasoma maculicauda* about 10 inches long (25 cm). This is a breeding male in nuptial dress. The fish was on display at the Nancy Aquarium in Nancy, France. It was photographed by Dr. Denis Terver.

Cichlasoma (Theraps) maculicauda

C. maculicauda has the widest distribution of all Central American cichlids, from the Rio Usumacinta in Mexico to the Rio Chagres in Panama. Its dispersal is restricted to the Atlantic drainage. The extensive distribution is reflected in the many variants and races known for this species. It is probably not its evolutionary success but rather its tolerance to sea water that assisted *C. maculicauda* in its dispersal. This species is observed regularly among marine fishes in pure sea water, hence it could easily spread via river estuaries to the south.

As could be anticipated, the mineral content of the aquarium water is of minor importance. The temperature may vary between 72° and 85° F (22-30°C). The maximum length of the male *C. maculicauda* lies around 30 cm (12 in), that of the female around 25 cm (10 in). At a size of about 13 cm (5 in) the fish start to stake

Cichlasoma maculicauda is the most widespread of all cichlids in Central America, probably because of its tolerance to sea water. Consequently it is found in many color varieties. This fish was photographed by Conkel.

Cichlasoma maculicauda photographed by Hans Mayland.

This male *Cichlasoma maculicauda* is a green variety photographed by the author, Ad Konings.

Above: An adult male 8 inches long (20 cm) on display at the Nancy Aquarium, Nancy, France. Photo by Dr. Denis Terver.

Left: This black belt (of karate fame) cichlid, *Cichlasoma maculicauda*, was photographed by Ken Lucas.

Cichlasoma maculicauda photographed at Steinhart Aquarium, San Francisco, California, by Dr. Herbert R. Axelrod.
Below: A pair of *Cichlasoma maculicauda* spawning. Photo by Rainer Stawikowski.

out their territories and should be given ample room to do so. When a pair forms you must remove all conspecifics from their quarters. Males can be nasty to each other. Male coloration is more pronounced and has more red markings than that of the female.

C. maculicauda is a notorious plant-eater, but due to its generalized diet it may thrive on crustaceans, fruits, and debris as well. In captivity this species can be satisfied easily with various sorts of food, but be careful of monotony. Alternate days of beefheart with days on which pelleted foods are offered.

The natural habitat of *C. maculicauda* is as diverse as its food. This species lives in stagnant water or in wide, slowly flowing streams. It likes to hide among rocks, logs, and plants, and it spawns in the open. In Lake Gatun, Panama, *C. maculicauda* was observed to hide its offspring in the rocky areas. This could be an adaptation to the large populations of introduced piscivores like *Cichla ocellaris*. The decor of the tank should contain some shelter for the female. A flat stone is used for spawning and as a visible territory marker.

Breeding in *C. maculicauda* may take place at an age of seven months and a size of about 15 cm (6 in). Although the male becomes territorial at this size, he

The three photographs on this page show a pair of *Cichlasoma maculicauda* spawning. The topmost photo is of a male. The center photo is of a female. Photos by Rainer Stawikowski.

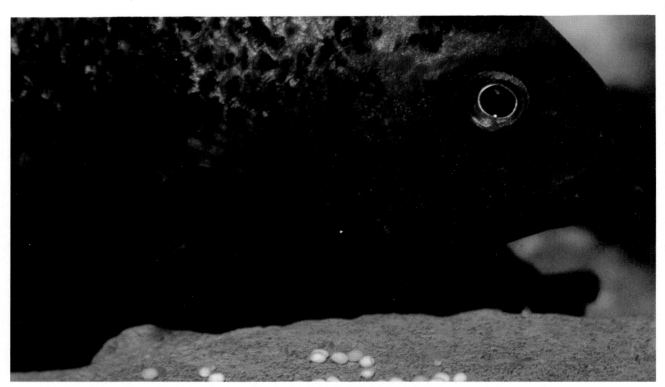

The female *Cichlasoma maculicauda* guarding her eggs, which have turned white and might very well be unfertilized. Usually the pair eat such eggs before they fungus. In the photo below, the pair guarding their eggs. Note that some of the eggs are white. These are infertile, dead eggs. The normal eggs are almost perfectly camouflaged. Photos by Rainer Stawikowski.

has not yet reached maturity. When breeding with siblings from the same brood, the first few spawns will not result in offspring. The first sign that breeding is at hand is the pair defending a territory. Shortly thereafter the female attains her nuptial color. The male acquires his breeding color much later or not at all, depending on the size and number of tankmates. The breeding colors are mainly dark, as is usual for open-water substrate-spawners. Especially the ventral part of the body darkens and the "black belt" gets wider and eventually turns the complete body black. The dorsal part of the head remains silvery and the nostril stripes become pronounced. Spawning is initiated by a thorough cleaning of the slab. Some pits are dug in the sand. A mature female *C. maculicauda* deposits over 1000 eggs that are laid in short rows. After every turn the male fertilizes the eggs just deposited. After 60 hours the eggs hatch and the larvae are released from the eggshells. The fry are mobile seven days after spawning and are guarded by both parents.

 C. maculicauda should not be kept together with *C. heterospilus*.

The female *Cichlasoma maculicauda* turns very dark when guarding her developing eggs. The white eggs are no good. The fertile, developing eggs are almost invisible. In the lower photo, one of the parent *C. maculicauda* tend to a mass of newly hatched fry that have not yet become free-swimming. Photo by Rainer Stawikowski.

After the eggs hatch the babies are moved to a depression in the sand where they are carefully protected. Photo by Rainer Stawikowski.

The fry are herded from one hiding place to another until they are free-swimming, as in the photograph below. The spawning series of *Cichlasoma maculicauda* was taken by Rainer Stawikowski.

Cichlasoma (Theraps) intermedium (GUENTHER, 1862)

Cichlasoma intermedium, a widespread cichlid found in Mexico, Guatemala and Belize. Photo by David Sands.

Cichlasoma (Theraps) intermedium

This species is found in Guatemala, Belize, and Mexico. It is restricted to clear and rather rapidly flowing water with a mineral content of between 300 and 600 microSiemens and a pH around 8.

The maximum size a male *C. intermedium* eventually may attain is not certain but must be at least 25 cm (10 in), probably similar to those of the other herbivorous *Theraps* (30 cm). Females remain somewhat smaller but are difficult to distinguish from the males. At a length of about 14 cm (6 in) for the male, this species may

Cichlasoma intermedium photographed by W. Heijns.

Cichlasoma intermedium. Photo by Konings.

Cichlasoma intermedium. Photo by Conkel.

Another *Cichlasoma intermedium* color variety. All of these color varieties exist in nature and are found in various parts of the range of this species. Photo by Rainer Stawikowski.

produce progeny. The young parents do not show a typical breeding dress. Larger females (20 cm, 8 in) become black ventrally and the markings on the sides intensify.

The natural food of *C. intermedium* is in general similar to that of the other herbivorous *Theraps*, but vegetable material is restricted to algae that form a fluffy layer over the stony bottom of the habitat. *C. intermedium* is frequently seen nibbling on the algae. In captivity any type of commercial pelleted food offered will be accepted. The most important factor in their well-being is clean and oxygen-rich water. The tank should contain some "obstacles" to break the artificial current produced by a powerful pump. If several females are kept with a single male, some rockwork should be added to provide them with shelter.

Breeding is imminent when the male has selected a female and both are cleaning a suitable spawning site. The site can be horizontal as well as vertical, according to the available material. A mature female may deposit over 1000 eggs that hatch in about three days. Five days after hatching the fry are mobile and are guarded by both parents. The female is much more responsible for the fry than the male, but the male acts very aggressively toward other tank inhabitants at the time the fry swim freely. Under natural circumstances the male may leave the female with her progeny a few days after the fry became mobile.

C. godmanni is a sibling species and is not recommended to be kept together with *C. intermedium.*

Cichlasoma (Theraps) hartwegi TAYLOR & MILLER, 1980

This cichlid, found only in Mexico, is called *Cichlasoma hartwegi*. Photo by Werner and Stawikowski.

Cichlasoma (Theraps) hartwegi

C. hartwegi is found in the streams and rivers that empty into the Presa de la Angostura in southern Mexico. In the Presa itself we may encounter *C. breidhori*.

The habitat of *C. hartwegi* may be of various types. These cichlids are observed in mountain streams and in stagnant pools. The water can be clear or turbid. The bottom may consist of rubble or of soft sediment. The mineral content is generally very high, and the pH ranges between 7.5 and 8.5. The temperature in the stagnant ponds may reach 85°F (30°C) without harm to the inhabitants.

C. hartwegi may grow to a maximum length of 30 cm (12 in) (males), but breeding may occur at a size of 15 cm (6 in). The sexes are difficult to distinguish, especially when dealing with fish from different broods. Males tend to be a little more colorful than females.

The natural food of *C. hartwegi* consists of anything edible, vegetable matter as well as invertebrates being eaten. In captivity they pose no problem in feeding, as any type of commercial food is taken. Care has to be taken not to pollute the aquarium water with the copious amounts of wastes these gluttons produce. Powerful filters and

frequent changing of the water should reduce the build-up of toxic components.

In order to breed this species you should supply them with vertically placed slabs or logs. The presence of unrelated species distracts the male's aggression. Breeding is initiated by staking out a territory and cleaning the spawning site. At first the female changes color. In fact, she loses most of her color and becomes black ventrally and very light, almost white, dorsally. After she has deposited the eggs, the male assumes a similar dress. The darkness of his breeding dress is dependent on the presence or absence of

A color variety of *Cichlasoma hartwegi.* Photo by the author, Ad Konings.

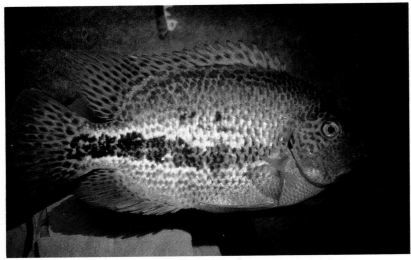

Another color variety of *Cichlasoma hartwegi.* Photo by Conkel.

intruding tankmates.

Three days after spawning the larvae are released from the shells and carried to safe quarters. Eight days after oviposition the fry are taken out of their shelter and they feed from the plankton. Brine shrimp nauplii are a satisfactory first food. You are asking for trouble when housing this species with *C. breidohri* or with a member of the following group.

Cichlasoma (Theraps) fenestratum (GUENTHER, 1860)

Cichlasoma fenestratum has a wide range and is very variable. Photo by Rainer Stawikowski.

Cichlasoma (Theraps) fenestratum

C. fenestratum belongs to the herbivorous group of the subgenus *Theraps*. Other members of the *fenestratum* group are *C. bifasciatum, C. guttulatum,* and *C. regani*. It is not advisable to keep two of the species mentioned above together in one tank. The whole group ranges over southern Mexico, but none of them have overlapping distribution patterns. Several of the species show geographical variation.

Cichlasoma fenestratum. Photo by Hans Mayland.

The water in which members of this group are encountered has a rather high mineral content (10 DH) and a pH between 7 and 8.5. The temperature lies between 72° and 82°F (22–28°C).

The maximum length *C. fenestratum* (and also the other species of the group) attains lies around 30 cm (12 in) for males and 25 cm (10 in) for females. Males usually have a more pronounced coloration than females. The final colors are not acquired before the fish have grown to 20 cm (8 in), but at a length of 14 cm (6 in)(males) the cichlids may successfully breed.

The natural food of these species consists of vegetable matter, but invertebrates are eaten as well. In captivity any commercial pellet food is readily accepted. Avoid feeding these gluttons soft and easily digestible food; in their greediness they may swallow excessively large morsels and choke. The natural habitat generally provides slowly flowing water on a soft bottom. Non-territorial individuals are normally encountered in midwater or close to the bottom in the middle of the stream.

For breeding purposes these cichlids need a log or a rock such as is usually found near the river banks. When breeding these beauties in captivity you should provide them with vertically placed slabs or logs. It is important to provide the breeding pair with the company of other, unrelated species. This will distract the aggression of the male from his consort. A shelter that will fit only the female is advisable.

Typical for *C. fenestratum* is the red coloration of the head and on the body behind the gill covers. This red coloration and the red in the fins disappear in both sexes when breeding is initiated. The final breeding coloration, attained when the fry become mobile, is a white body overlain with wide black bars. The ventral part of the fish, from the mouth to the tail, is completely black. A mature female may deposit more than 1000 eggs that are bigger than those of *C. synspilus*, hence their hatching time is double that of the latter species. Typically the female chooses a vertical rather than a horizontal place to spawn. Four days after spawning, the larvae are chewed from their eggshells and transported to a pit that was dug earlier. Twelve to 14 days after spawning the fry become mobile. They are considerably larger than those of *C. synspilus*.

Cichlasoma lentiginosum, from Mexico and Guatemala. These are two color variations. Photo above by Dr. W. Staeck; lower photo by W. Heijns.

Cichlasoma (Theraps) lentiginosum
(STEINDACHNER, 1864)

Cichlasoma lentiginosum, an ideal fish for the aquarium because it eats normal aquarium diet. Photo by W. Heijns.

Cichlasoma (Theraps) lentiginosum

This species from Mexico and Guatemala is a typical inhabitant of clear and flowing streams. The bottom of the typical stream is rocky and provides ample shelter and spawning sites. The water composition is not important for maintaining or breeding this beautiful species. Males may reach over 25 cm (10 in) in length, while the females remain a few centimeters smaller. Males are more colorful, but a clear distinction between the sexes can not be given.

The natural food of this cichlid consists of invertebrates, and it therefore is easy to keep in your tank. It accepts any type of food provided the food sinks to the bottom. Flake food may be offered and will be accepted after a short period of adaptation, but it is not recommended. Fry of this species can not be fed with floating dry food because they stay on the bottom. The decor of the aquarium should contain a cave into which both male and female can fit. It should be close to the bottom and have a rather flat ceiling.

Breeding is initiated by the female, who selects the spawning site. When breeding, both sexes lose most of their color and acquire a black and white pattern similar to that of other species in this subgenus: the entire body becomes white while the black markings intensify. It is interesting that the ventral part of the body becomes white instead of the black common in open-water substrate-breeders. The eggs are affixed to the ceiling of the cave and may number over 300. The female closely guards the eggs while the male watches outside the cave's entrance. After four days the eggs hatch and the larvae are assisted from the shells by the female. The larvae are kept on the bottom of the cave and become mobile 12 days after spawning. The parents take their offspring out of the cave to let them feed. The fry stay close to the floor and move around in the shadows of their parents.

C. coeruleus, C. irregulare, and *C. nebuliferum* are closely related and are not suitable as co-inhabitants of the community aquarium.

Cichlasoma (Theraps) nicaraguense (GUENTHER, 1864)

Cichlasoma nicaraguense, a beautiful male. Photo by Hans Joachim Richter. Below: *Cichlasoma nicaraguense.* Photo by Schmelzer.

Cichlasoma (Theraps) nicaraguense

Several variants of this species are known. The most commonly kept race comes from Costa Rican waters. The race from Lake Nicaragua displays a yellowish green color pattern. The natural waters are of moderate hardness, and the pH ranges between 7 and 8. The temperature fluctuates around 78°F (26°C).

The maximum length of the male is around 25 cm (10 in) whereas the female grows to a size of about 20 cm (8 in). The difference in the sexes is clearly

Above: An old *Cichlasoma nicaraguense* photographed at the Berlin (Germany) Aquarium by Dr. Herbert R. Axelrod.
Left: A female *Cichlasoma nicaraguense* photographed by Dave Tohir.

Cichlasoma nicaraguense, a male. Photo by Schmelzer.

small, elongated pit is dug between two stones. In the aquarium such sites are readily available. When kept with larger cichlids, *C. nicaraguense* may choose a real cave (when provided). The female deposits her eggs in the pit and they are fertilized by the male. Remarkably, the eggs (up to 400) are not attached to the substrate but are heaped in the spawning pit. The large pectoral fins of the female produce sufficient current to oxygenate the eggs in the pit. Eggs sprayed from the site by the current are collected and returned by both parents. The size of the eggs varies with that of the female. A mature female produces relatively large eggs (over 2 mm) that hatch after four days of incubation. The parents chew the larvae from the

distinguishable, in contrast to most other American cichlids. Females show a pronounced red coloration on the abdomen when maturity has been attained and an attractive blue color on the cheeks. Males gain a bronze-yellow coloration at a length of 8 cm (3½ in). At a size of 12 cm (5 in) they are able to breed.
The natural food of this descendant of the invertebrate-feeding *Theraps* consists of snails. This species forages from the bottom in the middle of Lake Nicaragua but moves inshore when breeding. Under artificial conditions *C. nicaraguense* eats anything offered and no special requirements are asked for. Inclusion of carotene-rich crustaceans (e.g., *Cyclops*) enhances the red color of the female.
While this cichlid forages from the flat and sandy lake floor, it breeds among rocks. However, it is not a real cave-breeder. It chooses a spawning site between the rocks of the transition zone, where a

A *Cichlasoma nicaraguense* photographed by Dr. Herbert R. Axelrod at the Steinhart Aquarium, San Francisco, California.

The eggs and emptied egg cases of an African Lake Tanganyika cichlid, *Neolamprologus,* are similar to those of *Cichlasoma nicaraguense.* Photo by Hans Joachim Richter.

eggshells. The larvae attach to the substrate since they possess typical adhesive glands on top of their heads. Twelve days after spawning the fry become mobile and are guarded by both parents. Due to the dense populations of *N. nematopus* in some Nicaraguan lakes, *C. nicaraguense* may fail to find a suitable spawning site in the rocky habitat. This may lead to frustrated males (probably males that lost their offspring) defending fry of another species. The other species, *C. dovii,* profits from these extra-specific foster fathers and is likely to produce more viable offspring. Mature *C. dovii* feed on *N. nematopus,* the breeding-site competitor of *C. nicaraguense.* Thus it seems that the latter species gets something in return for its outstanding efforts.

Chiclasoma nicaraguense photographed by Dr. W. Staeck.

Cichlasoma nicaraguense. Photo by H. P. Brock, Jr.

Cichlasoma (Theraps) sieboldii (KNER & STEINDACHNER, 1863)

Cichlasoma sieboldii. Photographed by the author, Ad Konings.

Cichlasoma (Theraps) sieboldii

C. sieboldii has its range on the Pacific side of Costa Rica and Panama. Two variants are known. This cichlid is a typical inhabitant of clear mountain streams. The mineral content of its natural waters is not high (up to 200 microSiemens) and the pH is around 7. The maximum size of the male is 20 cm (8 in). The female remains 5 cm (2 in) smaller. There is no clear difference between the sexes.

The biocover provides algae and invertebrates for *C. sieboldii* in the wild. In captivity they relish any prepared food except when it is soft and bulky. *C. sieboldii* is primarily herbivorous. The natural environment provides extensive rock formations in which this cichlid finds a spawning site. This cave-breeder can be accommodated with some stones piled up to create a dark recess in which they may readily spawn. Artificial caves, such as flowerpots, also are accepted. The initiative to breed comes

Cichlasoma sieboldii. Photo by Rainer Stawikowski.

from the female. She selects a site and attracts a male with her breeding colors. The non-breeding *C. sieboldii* displays a rather dull yellowish gray coloration, but the onset of breeding changes this pattern completely. In fact, the breeding colors resemble those of other cave-spawning *Theraps*: a white body with black markings. The eggs (100 to 300) are affixed to the ceiling or to the side of the cave. After three days the larvae are chewed free from the eggshells and collected inside the cave. Eight days after spawning, the first swimming fry can be seen venturing out of the cave. For the next few days the parents take their progeny out of the shelter and lead them to the foraging grounds.

A similar breeding behavior has been observed for *C. panamense*, a close relative of *C. sieboldii*.

Cichlasoma (Tomocichla) tuba MEEK, 1912

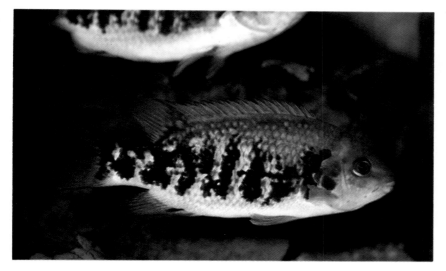

Cichlasoma tuba. Photo by Dr. W. Staeck.

Cichlasoma tuba. Photo by W. Heijns.

Cichlasoma tuba female guarding her spawn. Photo by W. Heijns.

Cichlasoma (Tomocichla) tuba

Notwithstanding the fact that some recent authors consider *C. tuba* to be the Atlantic sibling species of *C. sieboldii*, several important differences in their behavior are observed. *C. tuba* is encountered in the Atlantic drainage of Costa Rica and inhabits mainly mountain streams. The water conditions are unusual: a pH between 8 and 9 and a conductivity below 150 microSiemens. Nevertheless, *C. tuba* thrives in water with a higher mineral content (up to 800 microSiemens) and with a pH around 8.

In contrast to *C. sieboldii*, the maximum length of *C. tuba* lies around 30 cm (12 in) for males and 25 cm (10 in) for females. There is no clear difference between the sexes. In the wild, *C. tuba* is a herbivore and feeds on leaves and fruits fallen from overhanging trees. It is, however, not recommended to offer them half-rotten mangos or the like, because ordinary prepared pellet foods will do fine and will not mess up their quarters. Copious

Cichlasoma tuba. Photo by Endres.

amounts of beefheart and other soft and easy digestible foods mean trouble.

The natural habitat includes clear water and a rocky bottom, the water flowing rapidly and providing an oxygen-rich environment. In order to breed this interesting cichlid the water in the breeding tank should be well oxygenated. When breeding is at hand, the black markings are intensified, but a black cheek as in *C. sieboldii* is not present. Instead, the nostril stripes widen and give the appearance of a black mask. Remarkably, *C. tuba* is not always a cave-spawner like its sibling *C. sieboldii*, but spawns in open water on a horizontal slab, at least under artificial conditions.

Under natural conditions the availability of large caves and the pressure of large predators could induce this cichlid to breed in seclusion. It takes five days (at 80°F, 27°C) before the larvae have developed and are released from the eggshells.

Twelve to 14 days after spawning the fry become mobile. Brood sizes range from 200 to 500 fry. The fry are taken out into the moving water but remain close to the bottom.

Neetroplus nematopus GUENTHER, 1866

Neetroplus nematopus. This is a male in his normal garb. Photo by Hans Joachim Richter.

Neetroplus nematopus

N. nematopus is mainly a lacustrine species, although some individuals were observed in the river draining Lake Nicaragua, the Rio San Juan. The water chemistry is not important. *N. nematopus* not only is one of the smallest cichlids from Central America, but it is also one of the most successful.

Males may grow to a maximum size of 12 cm (5 in), but females remain considerably smaller, about 8 cm (3 in). Sexual differences are easily appreciated when dealing with individuals from the same brood, but there is no other distinguishing feature than the larger size of the male. It is best to start with about ten juveniles and let them grow up

together. Pair formation will occur at a length of about 6 cm (2½ in) (male). It is then better to net the other *N. nematopus* out of the tank unless they are housed in large tanks. This cichlid is rather aggressive and should be accompanied by much larger cichlids. Species like *C. citrinellum* or *C. longimanus* do fine with this active cave-spawner.

Neetroplus nematopus male guarding his free-swimming fry. The fry are free-swimming about 12 days after the eggs are laid. The male darkens up and looks almost like an African rift lake cichlid.

The head on this female *Neetroplus nematopus* seems to be developing a nuchal hump. Photo by David Sands.

This is a mature male *Neetroplus nematopus* in normal coloration. Photo by H. Ross Brock.

N. nematopus is the equivalent of an mbuna from Lake Nicaragua and is able to mow the entire algae-cover from the tank's decorations in a matter of days. When the stones are scraped clean *N. nematopus* accepts any other kind of prepared food. Pair bonds in Central American cichlids exist only during breeding, but this small representative may prolong the bond to the next spawning when no other ripe females are at hand.

Many individuals are found among the rocks in the Great Lakes of Nicaragua. The fact that this species also defends territories outside the breeding period (a feeding territory) makes it difficult for other species, like the closely related *C. nicaraguense*, to find a spawning site. Invariably, *N. nematopus* conquers any intruder on its territory, hence it is not advisable to accompany it with another cave-breeder.

The female initiates breeding by selection of the spawning-site and the display of the typical reversed color pattern. The eggs, about 150 for an adult female, are attached to the sides or ceiling of the cave. The relatively large eggs (2 mm) need four days to hatch. By the time the fry become mobile, about 12 days after spawning, the male has developed a black coloration similar to the female's. Both parents ferociously defend their progeny.

Cichlasoma (Archocentrus) nigrofasciatum
(GUENTHER, 1869)

One of aquarists' most popular Central American cichlids is the convict cichlid, *Cichlasoma nigrofasciatum*. It is very hardy and exists in a golden form also. Photo by Dr. Robert J. Goldstein.

Cichlasoma (Archocentrus) nigrofasciatum

C. nigrofasciatum is dispersed over a wide area and lives in the upper layer of rivers and brooks. It is encountered in almost any type of water and poses no maintenance problems in captivity. *C. nigrofasciatum* is among the most popular and hardiest cichlids known to hobbyists. The maximum size of an adult male does not exceed 14 cm (6 in), hence this species can be kept in small tanks (minimum 200 liters, 50 gallons). Females remain smaller (9 cm or 3½ in) and can be distinguished from the males by the reddish tinge on their abdomens. The red color is a sign of advancing maturity.

The natural diet of the convict cichlid is diverse but includes

This male convict cichlid, *Cichlasoma nigrofasciatum*, is shepherding his free-swimming fry. Photo by Hans Joachim Richter.

mostly invertebrates taken from the biocover. Copious amounts of low-nutrition algae are eaten as well, but only when it is not harvested by other species, like *N. nematopus* in Lake Nicaragua. As can be anticipated from its popularity among aquarists, feeding *C. nigrofasciatum* is simple.

The natural habitat supports numerous rocks usually larger than 30 cm in diameter. The spawning sites are dug in the sand under or between these rocks and barely fit the parents. The breeding colors are not much different from their casual dress. The black bars intensify and the background becomes whiter. The female sticks the eggs (about 100) to the ceiling of the cave. After three to four days the eggs hatch and the wrigglers are removed to a previously dug pit.

Top: One of the natural color forms of *Cichlasoma nigrofasciatum*. Photo by the author, Ad Konings.
Center: Another form of *Cichlasoma nigrofasciatum*. Photo by the author, Ad Konings.
Bottom: Young *Cichlasoma ornatum* are sometimes sold as a color variety of *Cichlasoma nigrofasciatum*. They are not color varieties of *C. nigrofasciatum*, however. Photo by W. Heijns.

Above: A fully mature male
Cichlasoma nigrofasciatum about
two years old. Photo by Ruda
Zukal.
Right: A golden variety of convict
cichlid was developed in Florida in
the early 1960's. It is doubtful that
they were natural fish found in
Central America. Photo by Mervin
Roberts.

Michael Gilroy, in Scotland, found this very interesting color variety. The young male *Cichlasoma nigrofasciatum* has a lot of rusty orange in its coloration, while the mature male has a lovely bright blue flush.

During the next five to six days the larvae are frequently removed to other pits. When the yolk is absorbed the fry rise from the bottom and are guided by the parents through the surroundings. When food runs short the fry may graze on the sides of their parents.

Intraspecific aggression in *C. nigrofasciatum* is not so pronounced as in many other Central American cichlids. Under natural conditions breeding territories can be contiguous and the dense populations of this successful cichlid may look like a

A male *Cichlasoma nigrofasciatum* protects its eggs inside the flowerpot "cave." This is a normal colored male. Photo by Ruda Zukal.

Nothing frightens the convict cichlid, *Cichlasoma nigrofasciatum*, from protecting its free-swimming fry. It even attacks a hand that is huge in comparison to the size of the fish. How many human parents would fight off an elephant threatening their young? Photo by Jaroslav Elias.

school. In Lake Nicaragua their territories are usually found among those of *C. citrinellum* at three to five meters (9-15 feet) depth. Fry and juveniles may be defended by several breeding pairs in the community.

Right: A color variety of *Cichlasoma nigrofasciatum.* Photo by Arend van den Nieuwenhuizen.

Below: Not all pairs spawn in caves; this pair spawns in the open on a rock. Photo by Ruda Zukal. Facing page: A pair of convict cichlids with their free-swimming fry. Photos like this one have made Hans Joachim Richter the world's leading aquarium fish photographer.

Cichlasoma (Archocentrus) centrarchus (GILL & BRANSFORD, 1877)

Cichlasoma centrarchus, a young male. The males seem much slimmer than the females, though this difference hasn't been mentioned in the literature. Photo by Dr. Herbert R. Axelrod.

The green variety of *Cichlasoma centrarchus*. Photo by the author, Ad Konings.

Cichlasoma (Archocentrus) centrarchus

C. *centrarchus* is a sibling species of *Herotilapia multispinosa*. The latter is specialized for an exclusively herbivorous diet, while *C. centrarchus* feeds predominantly on invertebrates (aquatic insects). *C. centrarchus* is found in very shallow and calm water. The habitat is densely vegetated and *C. centrarchus* finds spawning sites among the weeds. In contrast to other species of this subgenus, *C. centrarchus* is an open-water substrate-spawner. This is a consequence of living in the swampy shores of lakes and rivers where rocks and similar shelters are not available. Usually it spawns in very shallow water where predators are less frequent. *C. centrarchus* is also the largest *Archocentrus* and may reach a length of about 15 cm (6 in) (males). Immature specimens cannot be sexed easily. Males are larger than females, but this is evident only upon maturation.

Feeding *C. centrarchus* in captivity poses no problem. The species accepts any type of prepared pelleted food. The decoration of the tank should include a flat stone that could serve as a spawning site. Some aquatic plants are recommended, because this species hangs its offspring on the roots and stems of the plants.

In order to breed *C. centrarchus*, you could raise the temperature to 85°F (30°C), but this is not necessary. Shortly before depositing the eggs the female becomes almost completely black and resembles a breeding *H. multispinosa*. About 300 eggs are stuck to the substrate and are fanned by only the female. After three days the eggs hatch and are released from their eggshells by both parents. The larvae hang on roots, rocks, or the sides of the tank. The spot where the larvae are hung depends on the amount of oxygen in the water. In warm and stagnant water the larvae will be kept near the surface while in cooler or otherwise oxygen-rich water the larvae may be kept closer to the bottom. At night the whole brood is moved to a shelter close to the bottom and is hung out like laundry the next morning. After five days the larvae become free-swimming and follow both parents in the daily foraging routine.

A color variety of *Cichlasoma centrarchus*. Photo by W. Heijns.

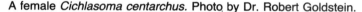

A female *Cichlasoma centrarchus*. Photo by Dr. Robert Goldstein.

A closeup of the head of *Cichlasoma centrarchus* showing the unique black markings on the edge of the operculum (gill cover). Photo by Dr. Herbert R. Axelrod.

Cichlasoma centrarchus is a sibling species of *Herotilapia multispinosa*. *Herotilapia* is a vegetarian while *C. centrarchus* thrives on aquatic insects. As with most living things, if the diet of preference is not available, the individuals adapt to that food that is readily available. Closeup photo of this head by Dr. Herbert R. Axelrod.

Herotilapia multispinosa (GUENTHER, 1866)

Herotilapia multispinosa spawning. The female is depositing eggs while the male stands by waiting to fertilize them. The eggs are fairly clear and thus take on the color of the substrate upon which they were deposited. Photo by Hans Joachim Richter.

Herotilapia multispinosa

Because evolution has taken the teeth of this species into consideration and developed them into optimum instruments to process algae and other vegetable material, this cichlid is classified in its own genus, but it has a close relationship with *C. centrarchus* and *C. spinosissimus*. *H. multispinosa* is distributed over most of Honduras and Nicaragua and is encountered in Pacific as well as Atlantic drainage systems. The water quality plays a minor role

Above: A breeding pair of the wild *Herotilapia multispinosa*. Note their intense yellow coloration. Photo by Dr. Herbert R. Axelrod. Below: Another pair of *Herotilapia multispinosa* spawning. Note the difference in coloration of the three breeding pairs on these two pages. Photo by Ruda Zukal.

This fish is being sold in petshops in America as *Herotilapia multispinosa*, but it probably is another species. Photo by Hans Mayland.

A very red female *Herotilapia multispinosa*. Photo by Hans Mayland.

Another fish being sold as *Herotilapia multispinosa*, but it is probably an impostor. Photo by Hans Mayland.

The green variety of *Herotilapia multispinosa.* Photo by Dr. Herbert R. Axelrod.

The black-tailed variety of *Herotilapia multispinosa*. Photo by Aaron Norman.

The dark color phase of *Herotilapia multispinosa*. This dark coloring is part of the breeding pattern as well as the pre-death pattern. Photo by Dr. Herbert R. Axelrod.

A pair of *Herotilapia multispinosa* preparing to spawn. Here they are cleaning the spawning site. Photo by Hans Joachim Richter.

Above: This male has assumed his dark color signifying he is herding his free-swimming fry. The colors of many cichlids change according to their creative phases. Photo by Hans Joachim Richter. Below: The female in the same color phase as the male above. Photo by Hans Joachim Richter.

since this species prefers backwaters and pools that remain after the water level has dropped in the dry season. Adult specimens become isolated in such pools on purpose and breed in an almost predator-free environment. Usually breeding *H. multispinosa* are found together with juvenile *C. managuense* of about 10 cm (4 in), which poses no threat to them.

H. multispinosa is one of the smallest cichlids from Central America. Males can attain a length of about 12 cm (5 in) and females 10 cm (4 in). There exist no obvious differences between the sexes. Fortunately this species is rather peaceful, and several males can be maintained in the same aquarium. The neutral color pattern of this cichlid includes a black spot on the side and one on the base of the tail. When a pair defend a territory the two spots join into a longitudinal stripe and the weak yellow background color turns into a lively golden yellow. The female prefers a vertical spawning site and deposits about 150 eggs. The eggs are guarded by both parents and hatch in three days. The larvae are hung onto roots and moved frequently. When after five days the fry begin to swim, the parents acquire an almost black coloration. The male shares care of the fry and actively defends his progeny.

Left: This amazing photograph shows the *Herotilapia multispinosa* fry being hung onto a plant located very low in the aquarium. Photo by Hans Joachim Richter. Facing page: A male *Herotilapia multispinosa* with his free-swimming fry. Photo by Hans Joachim Richter.

Cichlasoma (Amphilophus) longimanus
(GUENTHER, 1869)

Cichlasoma longimanus female. Photo by H. P. Brock, Jr.

Cichlasoma (Amphilophus) longimanus

C. *longimanus* has spread over a large part of Nicaragua and Costa Rica. Remarkably, this cichlid is found on both sides of the continental divide. At some places this divide is rather unclear and the two drainage areas (Pacific and Atlantic) are separated for just a few hundred meters. During this fish's evolution a section of a river system could have belonged to the other drainage area and could have taken its inhabitants along when its course was changed into the present state. The argument for this reasoning can be found in the color patterns and morphological features of *C. longimanus* found in these close but different drainages: they are very similar. Another noteworthy fact is the maximum size of

Cichlasoma longimanus male by Hans Mayland.

A young *Cichlasoma longimanus* male. Photo by Hans Joachim Richter.

The photographer identified this as a *Cichlasoma longimanus* and similar fish are sold as *longimanus,* but the long snout casts doubt on the identification. Photo by Conkel.

Another color variety of *Cichlasoma longimanus.* Photo by the author, Ad Konings.

Pacific and Atlantic races of this cichlid. The rivers that drain into the Caribbean are inhabited by much smaller (12 cm, 5 in) *C. longimanus* than those of the Pacific drainage region (maximum 18 cm, 7½ in). The reason is the availability of food.

Food is collected by filtering and screening the soft bottom of its biotope for invertebrates as well as vegetable matter. Muddy

develop against intruding conspecifics. Breeding occurs in a rocky habitat so *C. longimanus* can be regarded as a cave-breeder. (This breeding behavior was observed in the small variant of the species, and it is not known if the Pacific variants choose rocky protection as well.) As a true cave-breeder the female selects the spawning site and attracts the male. Eggs are

chew large food particles and spit the fragments in front of the fry.

A wild-caught male *Cichlasoma longimanus.* Photo by Dr. Herbert R. Axelrod.

bottoms usually are found in stagnant or slowly moving waters, the prevalent habitat for *C. longimanus*. When it is encountered with *C. rostratum*, the latter prefers the clearer and faster flowing part of the river while *C. longimanus* is pushed into the less oxygenated regions.

C. longimanus is rather peaceful and is frequently seen in large schools foraging and resting on the soft bottom. Only during breeding might some aggression

affixed to a vertical substrate and hatch in two days. It was observed in captivity that the larvae lacked functional adhesive glands on the head and fell to the substrate. This would mean that *C. longimanus* has not yet fully evolved into a cavity-spawner. Four days after hatching the fry become mobile but are still in a larval state as their yolk is not completely absorbed. Two days later the fry start eating. They grow rather slowly. The parents

Facing page: A closeup of the head of *Cichlasoma longimanus.* Note that the scales on the forehead run past the eye almost to the nostrils. The crater-like pores on the gill cover are also interesting. Photo by Dr. Herbert R. Axelrod.

Cichlasoma (Amphilophus) alfari MEEK, 1907

A young *Cichlasoma alfari* male. Photo by Uwe Werner.

Possibly a *Cichlasoma alfari* male. Photo by Hans Mayland.

Cichlasoma (Amphilophus) alfari

C. alfari is found in the Atlantic drainages of Nicaragua and Panama and in both Atlantic and Pacific drainages of Costa Rica. This species is recognized in many color variants and trophic races. Its habitat is in fast flowing rivers carrying clear water and with a rocky bed. Its food consists mainly of invertebrates that are picked from the hard substrate. *C. alfari* grows fast and is abundantly present in the waters of southern Costa Rica. Its great adaptability (many different morphological types are known) and fast growth rate make it a successful species that is in an active evolutionary state, producing new species when the circumstances are favorable. Closely related species are *C. diquis* and *C. rhytisma*, both from Costa Rica.

A terrible photo of a supposed *Cichlasoma alfari*. The fish seems to have been in shock and is showing a fright color pattern. This is verified by its lying on the bottom. Photo by Heiko Bleher.

Above: The female *Cichlasoma alfari* sifting sand through its gills. Below: The pair spawning. Photos by Uwe Werner.

A mature male *Cichlasoma alfari*. Photo by Uwe Werner.

As spawning time approaches for *Cichlasoma alfari*, the female begins digging pits for the future storage of her fry. Photo by Uwe Werner.

Like *C. longimanus, C. alfari* is not a real cave-breeder, as eggs can be deposited on a horizontal substrate as well. Usually the female selects the spawning site. The pair divide their parental tasks equally. The female is smaller (14 cm, 5½ in) than the male, which has a maximum size of 20 cm (8 in). *C. alfari* is rather peaceful toward conspecifics but may chase them when breeding. Breeding *C. alfari* lose much of their bright colors. The eggs hatch within four days and the larvae are moved to previously dug pits. Six days later fry become mobile and are led through the habitat in search of food. Sometimes one of the parents whirls debris from the bottom and it is then attentively screened by the agile fry.

The female *Cichlasoma alfari* with a batch of her fry that she has placed in one of the previously dug pits. Below: The pair with their free-swimming young. Photos by Uwe Werner.

Cichlasoma (Amphilophus) citrinellum
(GUENTHER, 1864)

A male *Cichlasoma citrinellum*. Photo by Hans Mayland.
Below: Old male *Cichlasoma citrinellum*. Photo by Dr. Herbert R. Axelrod.

Cichlasoma (Amphilophus) citrinellum

This species is distributed over a rather large area that spans the Atlantic drainages of Honduras, Nicaragua, and Costa Rica. Like many other *Cichlasoma*, *C. citrinellum* has been recorded in many different variants of color as well as morphology. This cichlid is popular in its yellow or orange color morph, but the most commonly encountered is the gray barred morph. The orange color morph is probably selected by its environment. The orange color is found in several species of *Parapetenia* (= *Nandopsis*) and in *Petenia* and is, as such,

A *Cichlasoma citrinellum* photographed at the Steinhart Aquarium by Ken Lucas. The bottom photo on the facing page was also taken at the Steinhart Aquarium in San Francisco.

This fish is a *Cichlasoma citrinellum* with many dark scales. Photo by Rainer Stawikowski.

Uwe Werner photographed this light orange male *Cichlasoma citrinellum.*

Below: Klaus Paysan photographed this light yellow *Cichlasoma citrinellum.*

This orange color variety is shown with its free-swimming fry. Photo by Dr. Herbert R. Axelrod. Below: Another color variety photographed by Dr. Herbert R. Axelrod.

present in the genetic material of probably all Central American cichlids. Two known populations of *C. citrinellum* do not contain orange individuals. In Lago Tiscapa it is the only cichlid observed and may therefore lack the orange variety. In heavily populated areas (Lake Nicaragua) spawning sites are scarce and many pairs are pushed to other areas to breed. It was observed that the orange-colored individuals predominantly bred in the deeper regions of the habitat. The dim light prevailing at

These facing pages show the most popular color varieties of *Cichlasoma citrinellum*.

The very light pink variety photographed by Ken Lucas.

The red variety with free-swimming fry. Photo by Hans Mayland.

The yellow-gold variety. Photo by Rainer Stawikowski.

The red-blotched variety. Photo by the author, Ad Konings.

The orange variety photographed by Hans Mayland.

The gray variety photographed by Rainer Stawikowski.

these depths may have selected the brightly colored morph because they are better seen by the fry, which feed on the flanks of their parents. At these depths a scanty food supply may enhance the selection for bright parents.

Another variation exists in the lips. Some populations contain individuals with fleshy lips. These populations are, however, never found together with *C. labiatum*, which is known for its over-developed lips.

Breeding takes place during the rainy season. Non-breeding *C. citrinellum* are found in large schools above any type of bottom. Breeding is preceded by pair-formation. Males acquire a pronounced nuchal hump and start displaying in front of females. When a pair is formed

Cichlasoma citrinellum spawning. Photo by H. P. Brock, Jr.
Below: The female *Cichlasoma citrinellum* mouthing the eggs she just laid. Photo by Rainer Stawikowski.

they try to secure a spawning site in the rocky habitat. If a suitable place cannot be found (it must be at least 2 meters (6 feet) in diameter) they may try their luck in other areas. *C. citrinellum* may even dig spawning pits in sandy areas. During the whole breeding period the pair rely on their fat reserves and will not eat until the fry are large enough to be abandoned. This is the reason why *C. citrinellum* breeds only during the rainy season, as during this period an overwhelming supply of insects and their larvae can condition the fish for breeding.

Most *Cichlasoma citrinellum* are gray with darker vertical barring. This color pattern becomes very distinct during spawning. Photo by Kenneth Thompson.

The gold *Cichlasoma citrinellum* with its free-swimming fry. Photo by Rainer Stawikowski.

Cichlasoma (Amphilophus) altifrons (KNER & STEINDACHNER, 1863)

Cichlasoma altifrons.

Cichlasoma (Amphilophus) altifrons

This cichlid belongs to the group of substrate-feeders. Observations in the field show that it is not restricted to sandy river beds but can also be found among pebbles and stones too big to chew. As a result of the coarse substrate, the lips of *C. altifrons* usually are thickened.

The distribution spans a rather large area, which is an indication of the fish's success. It is encountered in the Pacific drainages of Costa Rica, Panama, and Colombia. These waters are relatively poor in mineral content, unlike the majority of Central American streams. *C. altifrons* is commonly observed in moderately fast-moving rivers, where it can be found together with *C. sieboldii*.

This species feeds on invertebrates, which are probably not all filtered from the sandy bottom. In captivity this fish is often seen combing the biocover on stones. Not withstanding the similarity with cichlids of the

Above: *Cichlasoma altifrons* female. Below: *Cichlasoma altifrons* male collected and photographed by Heiko Bleher.

genus *Geophagus* from South America, *C. altifrons* developed in Central America and thus is a much more primitive member of the large cichlid family. The teeth on the pharyngeal bones are slender and thin at the margins and broad in the center, similar to other invertebrate-feeders. The teeth of some *Geophagus* are all a little thickened and better adapted to sustain the abrasion of the swallowed sand grains. The purpose of filtering sandy substrates is to extract the small food particles from the sand, which is accomplished much better by *Geophagus* than by *C. altifrons*. Nevertheless, this peculiar cichlid has adopted some breeding behavior from its South American counterparts. Like *Geophagus* (*Satanoperca*), *C. altifrons* conceals the eggs with a layer of sand.

Breeding is preceded by clearing a suitable spawning site. Both male and female may select the site and clean it. As is commonplace in Central American cichlids, the breeding colors are less attractive than the regular colors. After the eggs are

A female *Cichlasoma altifrons.* Photo by Conkel.

deposited and fertilized, both parents cover the eggs with sand. The pair alternate in guarding the spawn. Neither male nor female is found close to the concealed eggs, but they seem to purposely keep some distance away. After about four days the eggs hatch and the wrigglers are moved to a pit that was dug by one of the

parents. At this time the fish assume a darker color and the vertical bars become more pronounced. Seven days later the fry are mobile.

The maximum length of male *C. altifrons* is 24 cm (9½ in), females usually remain under 20 cm (8 in).

Like this *Geophagus australis, Cichlasoma altifrons* hides its eggs by covering them over with sand! Photo by Aaron Norman.

Cichlasoma (Thorichthys) aureum (GUENTHER, 1862)

Cichlasoma aureum female. Photographed by Dr. Herbert R. Axelrod at the Berlin Aquarium, Berlin, Germany.

Cichlasoma (Thorichthys) aureum

This *Thorichthys* is distributed in the Rio Motagua drainage around Amatique Bay. Its habitat consists of shallow water with soft or rocky bottoms. Although pair formation is only observed during breeding, many pairs can be seen defending their offspring in just a small section of the river. Territories are rather small and fry are abundant. *C. aureum* is found in many different waters with varying mineral contents.

C. aureum feeds from the bottom and filters the substrate. *Thorichthys* species are better adapted for foraging in shallow water than are *Amphilophus*. Species from the latter group feed in a more vertical position than do *Thorichthys*, which feed

Above: *Cichlasoma aureum* male. Photo by W. Heijns. Below: *Cichlasoma aureum* female. Photographed by the author, Ad Konings.

The top photo shows a young male yellow color variety of *Cichlasoma aureum*. The lower photo shows a very young male of the blue variety of *Cichlasoma aureum*. Both photographs by Rainer Stawikowski.

Above: Male *aureum* tending his eggs. Below: Young male gold *aureum* tending his eggs, too. Photos by Rainer Stawikowski.

in a more horizontal position. The natural food of *C. aureum* includes molluscs, although it is not specialized for such items. Comparison of the pharyngeal teeth of different *Thorichthys* species showed that those of *C. aureum* were the strongest and pointed to a diet of hard-shelled invertebrates. These reinforced teeth, however, are not of the same heavy type as those found in some *Parapetenia*. The general diet of *Thorichthys* contains insects and crustaceans, which usually are picked from the substrate or filtered from soft bottoms.

The maximum size of *C. aureum* lies around 15 cm (6 in) for the male and a bit smaller for the female.

During breeding a pair-bond is clearly noticed, but shortly before and after spawning there exists no bond whatsoever. This we have to bear in mind when breeding these cichlids in small tanks. The best remedy against aggression is to maintain a pair in a large (possibly community) tank. Before the actual spawning a pair is formed and together they clean a horizontal slab. The amber-colored eggs (about 300 per spawn) are guarded by male and female alternately. The rather small eggs hatch in two days and the larvae are transferred to a previously dug pit. During the following four days more pits are dug and the larvae changed from shelter to shelter accordingly. When fry become mobile they are taken out and led through the habitat, guarded by both parents. Fry of *Thorichthys* grow rather slowly and it may take a year before they are fully mature. During breeding the colors of both parents hardly change. Several color variants are known and imported. It is not clear if all variants are geographical races or if some among them belong to other species.

If you look closely, you can see a school of *aureum* fry being shepherded by their father. *Cichlasoma aureum* are excellent parents and it is simple to spawn them under the right conditions. Photos by Rainer Stawikowski.

Cichlasoma (Herichthys) carpinte (JORDAN & SNYDER, 1899)

Cichlasoma (Herichthys) carpinte

C. carpinte is distributed in the lower drainage region of the Rio Panuco in Central Mexico. As is common for the more primitive cichlids, many color variants are observed, and variation in body contours and pharyngeal teeth also are reported. This species lives at the edge of cichlid territory and demands rather little of its environment. In some of its waters the mineral contents exeed that of sea water. C. carpinte was also observed in pure sea water and can be found regularly in brackish water of coastal areas. In areas where it

Cichlasoma carpinte male. Photo by Hans Joachim Richter. Below: Another color variety of *carpinte*. This male was photographed by Klaus Paysan.

Young *Cichlasoma carpinte* male, a color variety. Photographed by Conkel.

This is a pool-raised *Cichlasoma carpinte*. It is the variety that is being commercially raised in Florida. This particular specimen won a prize in the Florida Tropical Fish Farms Association show and was photographed by Dr. Harry Grier.

Young *Cichlasoma carpinte* male. This is another color variety. Photo by Hans Mayland.

A female *Cichlasoma carpinte* tending her eggs. Notice that some eggs have been laid on the stone and additional eggs have been deposited on the glass. These are the eggs that have drawn the female's attention. Photo by Stanislav Frank.
Below: H. Hansen's photograph of an old male *Cichlasoma carpinte.*

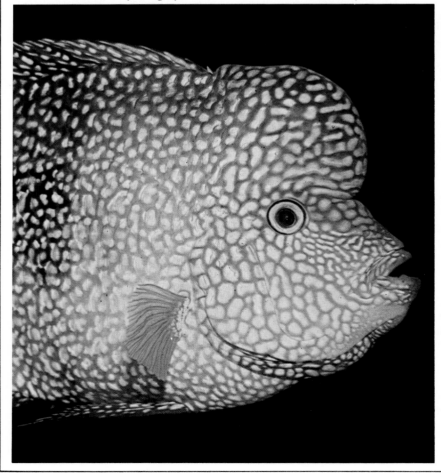

meets little or no other cichlids its maximum size can reach 30 cm (12 in), but where it finds competition from other cichlids (in the southern part of its distribution) its maximum size lies around 18 cm (7 in). The food consists of a wide variety of crustaceans, insects, snails, plants, and debris. When competition is present, *C. carpinte* seems to specialize more in vegetable matter. In regions where food is scarce this cichlid turns out to be aggressive toward conspecifics. This aggression is often experienced by aquarists who try to house a male and female in a small tank. If one of the two is not ready to breed, disaster may follow in a few hours. The best approach is to separate male and female with a divider in the tank. Both fish see each other but no harm can be done. At certain intervals you could remove the divider and watch carefully the reaction of the two fishes. If the female flees from the male the divider should be returned immediately. If a pair is placed in a large tank (over 500 liters or 125 gallons) and enough shelter is available, we would have better chances for a successful breeding.

Before spawning takes place a pair is formed and both male and female display similar coloration. When the eggs are deposited the female has a black color especially prominent on the lower and hind parts of her body. The eggs are affixed on a horizontal or vertical slab and hatch in three to four days. The larvae are moved to several pits that were dug around the spawning site. In five days the fry become mobile and the male acquires the typical breeding color and assists the female in defending offspring.

A color variety of *Cichlasoma carpinte*. This is a male. Photo by Hans Joachim Richter.

A young male color variant of *Cichlasoma carpinte*. Photo by Burkhard Kahl.

Another color variety of *Cichlasoma carpinte*. Photo by Conkel/Taylor.

THE ANCIENT CONTINENTS
Movement of the Continents.

Shown to the right is Pangaea after its breakup into the predecessors of the modern continents. The area above the line constitutes Laurasia; that below the line is Gondwana. The shading corresponds to the current continental limits as used in this book on Central American cichlids.

Notice that India, originally part of the southern continent, is now part of the northern continents, Southeast Asia of Eurasia. North and South America have undergone a complicated history of connection and disconnection across the Central American area, resulting in faunas that are remarkably different for such closely related areas. That's what makes this book so interesting!

When using these maps remember that the map of the modern continents is on a flat plane projection to allow all the areas to be seen in one surface while the map of the ancient continents is on a more polar projection, so distances are distorted. The smaller islands have been left off these maps for clarity.

THE MODERN CONTINENTS

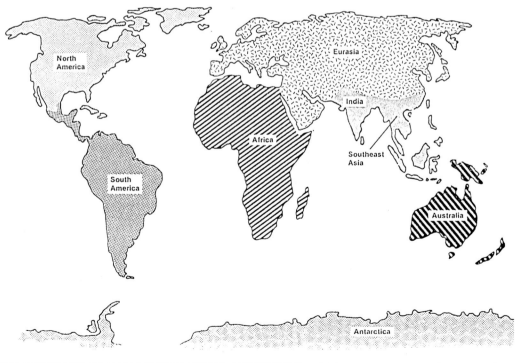

EVOLUTION OF CENTRAL AMERICAN CICHLIDS

Unquestionably, the cichlids from Africa and from South and Central America form a natural group. This implies common ancestry in Africa and a migration to the New World. It is tempting to presume that cichlids were already present when America and Africa were still united as part of the continent Gondwana, but paleontological data contradict the presence of cichlids at that epoch. The history of cichlids goes back about 50 million years. At that time, the beginning of the Oligocene, South America had already drifted quite far from Africa, complicating a suitable theory for the migration of cichlids.

Greenwood, in 1983, proposed a bridge of islands between Africa and South America with a concurrent low water level due to a severe Ice Age. This would have allowed secondary freshwater fishes as well as primates and rodents to migrate to the New World. A similar island bridge was proposed for the connection of South America with Nuclear Central America. Bussing, in 1976, assumed that a continuous land bridge connected South America with the northern continent and not an assembly of islands. The reasoning is that cichlids will not travel through open water but remain in close contact with the bottom—certainly true of the primitive members of the family. The drawback of an island bridge is the ocean around them. Inevitably, these avenues were too deep or too wide to cross. Following Bussing, other authors also seem to believe that "island hopping" is not to the liking of cichlids. The consequence is not only a continuous connection between North and South America, but also, in my opinion, between West Africa and South America 50 million years ago.

Nuclear Central America

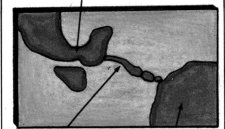

Island Bridge   **South America**

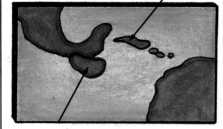

Honduras/Nicaragua

Cuba

Hispaniola

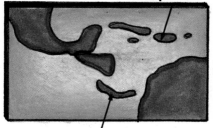

Costa Rica/Panama

The connection and disconnection of the island bridge between North and South America. To understand these maps you must read the text.

Although the two continents were rather far apart, West Africa could have been connected with the north of Brazil via the Atlantic Ridge. The fact that the Atlantic Ridge has the same contour as that of the continents adds credence to the feasibility of this hypothesis. According to this idea the distance between Africa and South America was exactly filled with the risen Atlantic Ridge, allowing primates, rodents, and fishes to populate the New World. A steep drop in the water level of the ocean may have aided this process.

It is thus likely that the first cichlids arrived in the north of today's Brazil, from which point they spread inside the continent.

The African cichlids may have belonged to two different genera, *Tylochromis* and *Hemichromis*, and entered primarily from West Africa. The streams of South America have a low mineral content leading to a poor nutritional base. The fierce competition was not the best start for the immigrant cichlids. The fact that the Amazon River was a lake before it found a drainage on the eastern side of the continent explains the number of species observed in this mighty river system. Nevertheless, the relative lack of food hampered the formation of the dense cichlid populations that are needed for rapid adaptation to a competitive environment. The coastal rivers and the lagoons contain more minerals, hence they offer more food and denser populations of aquatic life. It is here that cichlids may have thrived in their early days on the new continent.

Crossing close to shore from one river estuary to the next, they dispersed along the complete northern coast of South America. Even a few recent species from Central America are able to thrive in sea water. *C. maculicauda* was observed several times among marine fishes along the Caribbean shores of the continent. In a similar way and possibly shortly after their introduction into the New World, these African immigrants traveled along the assumed isthmus between South America and Nuclear Central America. Between 40 and 25 million years ago this isthmus was pushed eastward and broken up into the islands known today as the Antilles. Shortly after this isthmus moved eastward, a gap between the two America's was a fact. Not until three million years ago was this gap filled again with Costa Rica and Panama.

According to the geological timetable, there existed a period of at least 20 million years in which Nuclear Central America was completely isolated. We know that cichlids were present in Nuclear Central America at least 25 million years ago because a fossil cichlid, *Cichlasoma woodringi*, was discovered on Hispaniola (Haiti), one of the Caribbean islands.

Now that we have discussed briefly some geological background, the evolution of the Mesoamerican cichlids will be easier to explain. The evolution starts with a primitive cichlid belonging to the subgenus *Parapetenia* (= *Nandopsis*). This species found its way along the northern coast of South America and finally into Nuclear Central America. The evolution that takes its course in any population of living organisms is dependent on two factors: its physical environment and the interaction with other inhabitants of its biotope. The first factor acts very gradually and usually creates new species via geographical variants. The time span between the formation of a new species and its geographical isolation from its ancestor may be in the millions of years. The second factor may yield new species in just a fraction of that time. It involves adaptation in the total fish-community of a river system. Such adaptations are only needed when the species is suddenly confronted with another species with the same food requirements. The resulting competition forces both species to specialize on a different part of the previous mutual diet.

Before the complete impact of these two factors can be appreciated we have to define a

Cichlasoma maculicauda. Photo by Hans Mayland.

species. What is a species? Not so long ago a species was thought to be a population of anatomically identical organisms. The individuals of such a population could be interbred and the progeny were to have been identical in appearance and fertile. The infertility of artificially bred offspring from parents that clearly belonged to two different species seemed to justify this rule. Later it was observed that offspring from a cross between rather distinct species were fertile. From then on a new definition of a species had to be found. Up to now there is no such clearly defined description of the species, at least not for cichlids. The definition that most scientist will agree upon is the following. *A species is a group of organisms that recognize each other as belonging to the same group and mate unselectively.* The recognition should be under natural conditions and only those individuals that share their genetic material with the whole pool of the population belong to that species. This is very important because it limits the variability of the species. As long as the population consists of a fluently dispersed group, it is not difficult to determine a species. When the distribution has a disjunct character it becomes more troublesome. Taxonomy usually is not a favorite subject for cichlid enthusiasts, but frequent changing of scientific names may disturb them.

Greenwood and Kullander are probably the first scientists who described species according the above mentioned definition. This could imply that closely related species are the result of splitting an ancestral species into two geographically isolated populations. When the isolation is long enough that the two populations do not recognize each other when brought back together, we should correctly speak of two species. Unfortunately, these recognition experiments are virtually impossible to perform under natural conditions, so ichthyologists have to formulate other definitions. Usually they combine the geological history of the region with anatomical and behavioral data on the fishes to describe closely related species.

Thus the theoretical side of a species. Before I discuss speciation I would like to point to the following frequently overlooked biological fact: from the total number of fry one pair of cichlids (or other organisms) produce during their entire life, on the average only TWO will

eventually mature and replace their parents in the population. All other offspring, sometimes numbering in the thousands, will fall prey to the environment. For those who think that from a successful species a few more may survive, I give the following calculation. Imagine one pair of *C. managuense* in Lake Nicaragua. This lake has a surface area of 3000 sq. miles (8000 sq. km) and has an average depth of 40 feet (12 m). The volume of this large lake thus measures 96,000,000,000,000 (96,000 trillion) liters of water. A mature pair of *C. managuense* must have a volume of at least one liter. Assume further that this pair will spawn only twice in their life and produce 1000 wrigglers in

Below: *Cichlasoma managuense*. This is a male about 10 inches (25 cm) long. Photo by Dr. Denis Terver at the Nancy Aquarium, France. On the facing page we see a photo sequence of a female *Cichlasoma managuense*(?) as she attacks a swordtail, *Xiphophorus helleri*. Her mouth extends and she sucks in the swordtail, swallowing 2.5 cm (an inch) immediately. She holds the fish in her mouth until it stops moving and then swallows it in three gulps. Photos by Hans Joachim Richter.

total. If there is no loss of fry, these 1000 will be mature in two years and require a volume of 500 liters. These 1000 mature *C. managuense* form 500 pairs and each produce another 1000 fry in the next two years, yielding 500,000 adult cichlids in four years. If we continue in a similar fashion the lake will be filled to the brim with *C. managuense* in just 10 years! Admittedly, the survival of all fry is not feasible even under artificial conditions, so let us also calculate what happens when three instead of just two larvae grow to full maturity. This means three adult *C. managuense* after two years and ten in eight years. If we continue we will find the lake completely crammed in 153 years. Some cichlid communities may exist for millions of years. Seen in this context 153 years means nothing. It must be clear that only two descendants of thousands of fry born from one pair of cichlids reach maturity, not even three.

I gave this example not only to show the impact of the devastating selection nature takes, but also to indicate the massive number of cichlids born. It is the number of cichlids that defines the variability of the species. The occurrence of an advanced mutation in the genetic material increases with the number of offspring. If a fry is born with a mutation that gives it an infinitely small but better chance to survive, it probably will survive and may be able to pass on its mutant gene to its own offspring. According to calculations, such a small advantage may be integrated into the population's gene pool in a few years!

Most mutations either are not selective or act against the survival of the affected fry, but the vast number of fry born over the long period a species may exist explains why a population is always optimally adapted to its environment. Changes will not occur unless a new selective mechanism enters its habitat.

The color of cichlids is an important feature for species recognition, but color itself is rather irrelevant to the survival of the individual. Color is totally accidental, but conspicuous colors may be selected against by predators. Although color is, within certain margins, not relevant to a particular population, it is very important in speciation. The following example will clarify this. Imagine one species dispersed through a single river system. There is not an isolated individual and all share in the total pool of inheritable material (there is free mating among all members of the population). If a young is born with an aberrant color and even if it grows to maturity, it will not be recognized by the other members of the population. Thus it cannot propagate, leading to the loss of its inheritable material. One day a volcano erupts and cuts off an upstream section of the river system, creating an isolated lake. The cichlids in the downstream section have no contact with the population in the lake. Time passes and both populations develop differently colored variants of the previous species. This may take tens of thousands of years. This process is evolution caused by accidental and minor changes in the population. The diet of both populations, however, remains exactly the same. The same volcano blasts its hot lava over the lake again and the water finds its way around it. Now the lake population enters the downstream part of the river and meets the resident cichlids. The time of separation has lasted long enough to hinder the recognition of either of the two geographical variants as belonging to one species. Now each population sees the other as a food competitor because both are dining from the same ecological niche. This immediately triggers the selection of any mutant offspring that taps a slightly different food source. The strong selection favors such mutants and (see calculation) in a matter of a few hundred years two species are inhabiting the river system.

Now we are going to apply this to the cichlid fauna of Central America. *C. haitiensis* and *C. vombergae* are the only two species encountered on Hispaniola. These two species belong to *Parapetenia*, which harbors the most *primitive* species. By *primitive* is meant that the species may show a substantial variety in its diet. The central mountain ridge of Hispaniola separates the river systems into two drainages. In one drainage area (Haiti) *C. haitiensis* is observed, whereas the other area (Dominican Republic) harbors *C. vombergae*. Both resemble each other to a great extent. Nevertheless, these two species may have been separated for over 25 million years! Moreover, *C. tetracanthus* from Cuba, isolated from *C. haitiensis* for at least the same period of time, can hardly be told apart from the former two. *C. woodringi*, the 25-million-year-old fossil cichlid, resembles these cichlids in detail.

Lake Apoyo in Nicaragua has an estimated age of about 10,000 years. This, remarkably, was long enough to create a new species, *C. zaliosum*, because its ancestor was displaced by *C. citrinellum* and was forced to specialize. If

10,000 years is long enough to make a new species, a mere 2500 species could theoretically inhabit Cuban waters! This illustrates the difference between geographical evolution and competition evolution.

The cichlids on Cuba, *C. tetracanthus* and *C. ramsdeni*, were always isolated and show the maximum difference after 25 million years of isolation. For *C. tetracanthus* several subspecies were described. Most of them were based on age-related anatomical differences, while others were geographical variants. If these geographical variants do not recognize each other when brought together, they comprise different species. Modern ichthyologists regard all of them as variants within one species. *C. ramsdeni* experienced a much better isolation (until recent human intervention) and is regarded as a true species. These two species never suffered competition from similar species and were never forced to specialize. *C. ramsdeni* was purposely spread over the other drainage regions of Cuba to save it from extinction. This undoubtedly triggered the feeding specialization of both species. That it will result in a new species is not necessary, but it may be expected.

Another area where cichlids suffered relatively little competition is the northwestern corner of South America. *Parapetenia* is observed here together with *Aequidens*. When the isthmus between the two Americas moved eastward, cichlids remained at the northern coast of South America. They endured relatively little competition and remained in a fairly primitive state. During the last two million years, new species entered their territory from the north and at least one

This is a fairly rare specimen of Central American cichlid, *Cichlasoma zaliosum*. Photo by Dr. Paul Loiselle.

species, *C. atromaculatum*, must have adapted itself to the immigrants.

Regions with outstanding speciation are encountered at the borderline between two previously isolated areas. This is observed in and around Lake Nicaragua, around Bahia de Amatique, and around the Isthmus of Tehuantepec. The relatively small biogeographical region Yucatan/Guatemala is bordered on both sides by such margins. This is clearly illustrated by the number of different cichlid species found in this region: 38! The region Central Mexico harbors only 13 different cichlid species, of which eight belong to the primitive subgenus *Parapetenia*. Of the 38 species of Yucatan/Guatemala only five belong to this subgenus, clearly indicating the advanced state of that cichlid community. The biogeographical region Honduras/Nicaragua is bordered

by one confrontation area and a lake and harbors 25 species. The relatively new region, Costa Rica/Panama, was mainly populated from the north but still supports 20 different species (the species in Rio San Juan not counted) in its waters. This region was populated by already specialized cichlids. This is reflected in the number of species belonging to *Parapetenia*: only two, *C. dovii* and *C. umbriferum*! These two are the largest *Parapetenia* known and are specialized piscivores.

Again I would like to stress the fact that any population of living organisms is optimally built to live in its environment. Up to a few hundred years ago, evolution took its course by a very gradual change in inheritable characteristics and by rapid competitive events, elicited by geological processes. Artificial dispersal by humans unwillingly provokes speciation at a much faster speed.

The illustrations on these pages may, in some cases, duplicate the color photographs already appearing in the text. They have been repeated in the Illustrated Atlas to make it handier for the reader. An attempt was made to make the Illustrated Atlas in alphabetical order but the constant changes in Central American cichlid nomenclature, much of which is by amateur ichthyologists and thus very suspicious, has made the task impossible. A number of South American cichlid species also have been included in this Atlas.

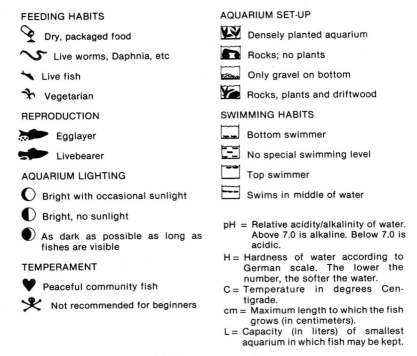

SYMBOLS

FEEDING HABITS		AQUARIUM SET-UP	
	Dry, packaged food		Densely planted aquarium
	Live worms, Daphnia, etc		Rocks; no plants
	Live fish		Only gravel on bottom
	Vegetarian		Rocks, plants and driftwood

REPRODUCTION — Egglayer / Livebearer

SWIMMING HABITS — Bottom swimmer / No special swimming level / Top swimmer / Swims in middle of water

AQUARIUM LIGHTING — Bright with occasional sunlight / Bright, no sunlight / As dark as possible as long as fishes are visible

pH = Relative acidity/alkalinity of water. Above 7.0 is alkaline. Below 7.0 is acidic.
H = Hardness of water according to German scale. The lower the number, the softer the water.
C = Temperature in degrees Centigrade.
cm = Maximum length to which the fish grows (in centimeters).
L = Capacity (in liters) of smallest aquarium in which fish may be kept.

TEMPERAMENT — Peaceful community fish / Not recommended for beginners

Photography

This Atlas section contains photos that are the work of many different photographers listed here.

Hiromitsu Akiyama
K. Attwood
Heiko Bleher
Dr. Martin Brittan
Dr. Warren E. Burgess
Dr. Brooks Burr
Vojtech Elek
Dr. Augustin Fernandez-Yepez
Walter Foersch
Dr. Stanislav Frank
H. J. Franke
Dan Fromm
Dr. Robert J. Goldstein
Dr. Myron Gordon
Dr. Harry Grier
K. Jeno
Burkhard Kahl

S. Kochetov
R. Lawrence
Ken Lucas, Steinhart Aquarium
Gerhard Marcuse
Dr. Richard L. Mayden
Hans Mayland
Manfred Meyer
Midori Shobo (Fish Magazine, Japan)
Leo G. Nico
Aaron Norman
Dr. Lawrence M. Page
Klaus Paysan
Kurt Quitschau
Hans-Joachim Richter
Mervin F. Roberts
Erhard Roloff

Andre Roth
Jorgen Scheel
Gunter Schmida
Harald Schultz
Dr. Wolfgang Staeck
Rainier Stawikowski
Donald C. Taphorn
Edward C. Taylor
Dr. D. Terver, Nancy Aquarium
Gerald J. M. Timmerman
Arend van den Nieuwenhuizen
Braz Walker
Franz Werner
Uwe Werner
Gene Wolfsheimer
Ruda Zukal

Aequidens pulcher pH7;H10;24C;20cm;150L ∿ ➤ ◐ ✕ ▧ ▣

Aequidens pulcher pH7;H10;24C;20cm;150L ∿ ➤ ◐ ✕ ▧ ▣

Aequidens itanyi pH7;H12;26C;14cm;100L ♀ ➤ ◐ ♥ ▧ ▣

Aequidens coeruleopunctatus pH7;H8;25C;15cm;100L ∿ ➤ ◐

Aequidens "silver edge" pH7;H10;25C;15cm;200L ♀ ➤ ◐ ✕

Aequidens "silver edge" pH7;H10;25C;15cm;200L ♀ ➤ ◐ ✕ ▧ ▣

Aequidens rivulatus pH7;H10;25C;16cm;200L ♀ ➤ ◐ ✕ ▧ ▣

Aequidens rivulatus pH7;H10;25C;16cm;200L ♀ ➤ ◐ ✕ ▧ ▣

Cichlasoma affinis pH8;H10;25C;14cm;200L 〜 ➤ ◑ ♥ 🖼 ⊟   *Cichlasoma affinis* pH8;H10;25C;14cm;200L 〜 ➤ ◑ ♥ 🖼 ⊟

Cichlasoma alfari pH7;H3;25C;20cm;400L 〜 ➤ ◑ ✕ 🖼 ⊟   *Cichlasoma alfari* pH7;H3;25C;20cm;400L 〜 ➤ ◑ ✕ 🖼 ⊟

Cichlasoma alfari pH7;H3;25C;20cm;400L 〜 ➤ ◑ ✕ 🖼 ⊟   *Cichlasoma altifrons* pH7;H3;25C;24cm;400L 〜 ➤ ◑ ✕ 🖼 ⊟

Cichlasoma altifrons pH7;H3;25C;24cm;400L 〜 ➤ ◑ ✕ 🖼 ⊟   *Cichlasoma altifrons* pH7;H3;25C;24cm;400L 〜 ➤ ◑ ✕ 🖼 ⊟

Cichlasoma labridens pH7;H3;25C;25cm;600L ♀ ➤ ◑ ✕ ▨ ▣

Cichlasoma alfari pH7;H3;25C;20cm;400L ∿ ➤ ◑ ✕ ▨ ▣

Cichlasoma alfari pH7;H3;25C;20cm;400L ∿ ➤ ◑ ✕ ▨ ▣

Cichlasoma aureum pH7;H3;25C;14cm;300L ∿ ➤ ◑ ✕ ▨ ▣

Cichlasoma bifasciatum pH7;H3;25C;30cm;800L ♀ ➤ ◑ ✕ ▨ ▣

Cichlasoma carpinte pH7;H3;25C;20cm;500L ♀ ➤ ◑ ✕ ▨ ▣

Cichlasoma citrinellum pH7;H3;25C;30cm;500L ♀ ➤ ◑ ✕ ▨ ▣

Cichlasoma citrinellum pH7;H3;25C;30cm;500L ♀ ➤ ◑ ✕ ▨ ▣

Cichlasoma macracanthum pH7;H5;25C;25cm;600L ♀ 🐟 ◑ ✂ 🖼 🔲   *Cichlasoma macracanthum* pH7;H5;25C;25cm;600L ♀ 🐟 ◑ ✂ 🖼 🔲

Cichlasoma robertsoni pH7;H3;25C;20cm;400L ♀ 🐟 ◑ ✂ 🖼 🔲   *Cichlasoma altifrons* pH7;H3;25C;24cm;400L 〰 🐟 ◑ ✂ 🖼 🔲

Cichlasoma alfari pH7;H3;25C;20cm;400L 〰 🐟 ◑ ✂ 🖼 🔲   *Cichlasoma alfari* pH7;H3;25C;20cm;400L 〰 🐟 ◑ ✂ 🖼 🔲

Cichlasoma altifrons pH7;H3;25C;20cm;400L 〰 🐟 ◑ ✂ 🖼 🔲   *Cichlasoma* cf. *labiatum* pH7;H3;25C;25cm;600L ♀ 🐟 ◑ ✂ 🖼 🔲

Cichlasoma festae pH7;H10;25C;50cm;800L ♀ ➤ ◑ ✕ 🎞 ⊡

Cichlasoma tetracanthus pH7;H3;25C;25cm;400L ♀ ➤ ◑ ✕

Cichlasoma carpinte pH7;H3;25C;20cm;500L ♀ ➤ ◑ ✕ 🎞 ⊡

Cichlasoma carpinte pH7;H3;25C;20cm;500L ♀ ➤ ◑ ✕ 🎞 ⊡

Cichlasoma cyanoguttatum pH7;H3;23C;25cm;500L ♀ ➤ ◑ ✕ 🎞 ⊡

Cichlasoma atromaculatum pH7;H3;25C;24cm;500L ♀ ➤ ◑ ✕

Cichlasoma atromaculatum pH7;H3;25C;24cm;500L ♀ ➤ ◑ ✕ 🎞 ⊡

Cichlasoma atromaculatum pH7;H3;25C;24cm;500L ♀ ➤ ◑ ✕

181

Cichlasoma atromaculatum pH7;H3;25C;24cm;500L ♀ 🐟 ◗ ✕ 🖼 ⊟   *Cichlasoma atromaculatum* pH7;H3;25C;24cm;500L ♀ 🐟 ◗ ✕ 🖼 ⊟

Cichlasoma aureum pH7;H3;25C;14cm;300L 〜 🐟 ◗ ✕ 🖼 ⊟   *Cichlasoma aureum* pH7;H3;25C;14cm;300L 〜 🐟 ◗ ✕ 🖼 ⊟

Cichlasoma bartoni pH7;H3;25C;18cm;500L ♀ 🐟 ◗ ✕ 🖼 ⊟   *Cichlasoma bartoni* pH7;H3;25C;18cm;500L ♀ 🐟 ◗ ✕ 🖼 ⊟

Cichlasoma beani pH7;H3;25C;30cm;800L ♀ 🐟 ◗ ✕ 🖼 ⊟   *Cichlasoma bifasciatum* pH7;H3;25C;30cm;800L ♀ 🐟 ◗ ✕ 🖼 ⊟

Cichlasoma meeki pH7;H5;25C;12cm;100L ♀ 🐟 ◑ ✂ 🖼 ⊡   *Cichlasoma meeki* pH7;H5;25C;12cm;100L ♀ 🐟 ◑ ✂ 🖼 ⊡

Cichlasoma aureum pH7;H3;25C;14cm;300L 〜 🐟 ◑ ✂ 🖼 ⊡   *Cichlasoma aureum* pH7;H3;25C;14cm;300L 〜 🐟 ◑ ✂ 🖼 ⊡

Cichlasoma bartoni pH7;H3;25C;18cm;500L ♀ 🐟 ◑ ✂ 🖼 ⊡   *Cichlasoma bartoni* pH7;H3;25C;18cm;500L ♀ 🐟 ◑ ✂ 🖼 ⊡

Cichlasoma psittacus pH6;H1;26C;30cm;500L 〜 🐟 ◑ ♥ 🖼 ⊡   *Cichlasoma psittacus* pH6;H1;26C;30cm;500L 〜 🐟 ◑ ♥ 🖼 ⊡

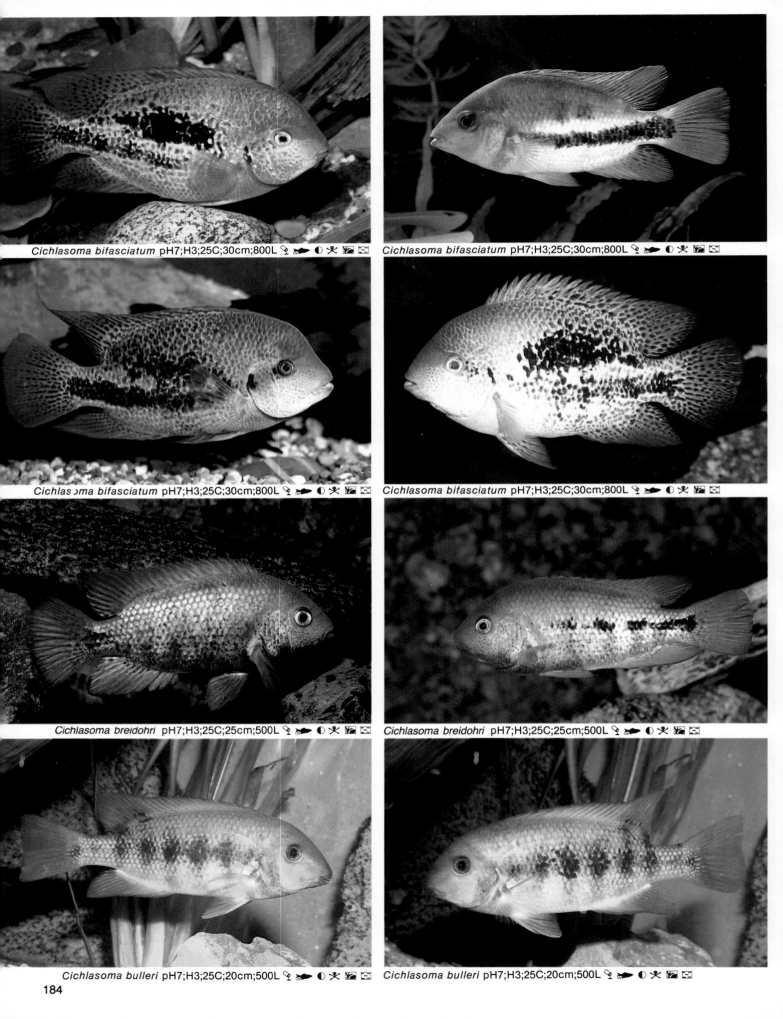

Cichlasoma bifasciatum pH7;H3;25C;30cm;800L ♀ ➤ ◑ ✕ ▨ ▭   *Cichlasoma bifasciatum* pH7;H3;25C;30cm;800L ♀ ➤ ◑ ✕ ▨ ▭

Cichlasoma bifasciatum pH7;H3;25C;30cm;800L ♀ ➤ ◑ ✕ ▨ ▭   *Cichlasoma bifasciatum* pH7;H3;25C;30cm;800L ♀ ➤ ◑ ✕ ▨ ▭

Cichlasoma breidohri pH7;H3;25C;25cm;500L ♀ ➤ ◑ ✕ ▨ ▭   *Cichlasoma breidohri* pH7;H3;25C;25cm;500L ♀ ➤ ◑ ✕ ▨ ▭

Cichlasoma bulleri pH7;H3;25C;20cm;500L ♀ ➤ ◑ ✕ ▨ ▭   *Cichlasoma bulleri* pH7;H3;25C;20cm;500L ♀ ➤ ◑ ✕ ▨ ▭

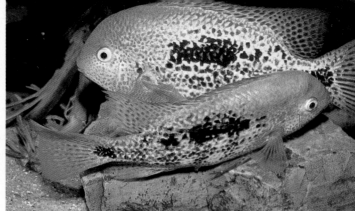

Cichlasoma bifasciatum pH7;H3;25C;30cm;800L ♀ 🐟 ◑ ✖ 🎞 🖼

Cichlasoma bifasciatum pH7;H3;25C;30cm;800L ♀ 🐟 ◑ ✖ 🎞 🖼

Cichlasoma bifasciatum pH7;H3;25C;30cm;800L ♀ 🐟 ◑ ✖ 🎞 🖼

Cichlasoma bifasciatum pH7;H3;25C;30cm;800L ♀ 🐟 ◑ ✖ 🎞 🖼

Cichlasoma guttulatum pH7;H5;25C;25cm;400L ♀ 🐟 ◑ ✖ 🎞 🖼

Cichlasoma guttulatum pH7;H5;25C;25cm;400L ♀ 🐟 ◑ ✖ 🎞 🖼

Cichlasoma intermedium pH7;H15;25C;25cm;400L ♀ 🐟 ◑ ✖ 🎞 🖼

Cichlasoma cf. *guttulatum* pH7;H5;25C;25cm;400L ♀ 🐟 ◑ ✖ 🎞 🖼

Cichlasoma bimaculatum pH6;H1;28C;15cm;150L ♀ 🐟 ◐ ♥ 🖼 ⊡

Cichlasoma bimaculatum pH6;H1;28C;15cm;150L ♀ 🐟 ◐ ♥ 🖼 ⊡

Cichlasoma coryphaenoides pH5;H1;28C;25cm;400L 〰 🐟 ◐ ♥

Cichlasoma coryphaenoides pH5;H1;28C;25cm;400L 〰 🐟 ◐ ♥ 🖼 ⊡

Cichlasoma coryphaenoides pH5;H1;28C;25cm;400L 〰 🐟 ◐ ♥

Cichlasoma facetum pH6;H1;24C;15cm;100L ♀ 🐟 ◐ ♥ 🖼 ⊡

Cichlasoma facetum pH6;H1;24C;15cm;100L ♀ 🐟 ◐ ♥ 🖼 ⊡

Cichlasoma facetum pH6;H1;24C;15cm;100L ♀ 🐟 ◐ ♥ 🖼 ⊡

186

Cichlasoma calobrense pH7;H3;25C;25cm;600L 〰 ➤ ◑ ✕ 🖼 ⊟   *Cichlasoma calobrense* pH7;H3;25C;25cm;600L 〰 ➤ ◑ ✕ 🖼 ⊟

Cichlasoma carpinte pH7;H3;25C;20cm;500L ⚲ ➤ ◑ ✕ 🖼 ⊟   *Cichlasoma carpinte* pH7;H3;25C;20cm;500L ⚲ ➤ ◑ ✕ 🖼 ⊟

Cichlasoma carpinte pH7;H3;25C;20cm;500L ⚲ ➤ ◑ ✕ 🖼 ⊟   *Cichlasoma carpinte* pH7;H3;25C;20cm;500L ⚲ ➤ ◑ ✕ 🖼 ⊟

Cichlasoma centrarchus pH7;H3;25C;15cm;300L ⚲ ➤ ◑ ✕ 🖼 ⊟   *Cichlasoma centrarchus* pH7;H3;25C;15cm;300L ⚲ ➤ ◑ ✕ 🖼 ⊟

187

Cichlasoma citrinellum pH7;H3;25C;30cm;500L ♀ ➤ ◑ ✄ 🖬 🖾   Cichlasoma citrinellum pH7;H3;25C;30cm;500L ♀ ➤ ◑ ✄ 🖬 🖾

Cichlasoma coeruleus pH7;H24;25C;13cm;200L ∿ ➤ ◑ ✄ 🖬 🖾   Cichlasoma coeruleus pH7;H24;25C;13cm;200L ∿ ➤ ◑ ✄ 🖬 🖾

Cichlasoma cyanoguttatum pH7;H3;23C;25cm;500L ♀ ➤ ◑ ✄ 🖬 🖾   Cichlasoma cyanoguttatum pH7;H3;23C;25cm;500L ♀ ➤ ◑ ✄ 🖬 🖾

Cichlasoma aff. cyanoguttatum pH7;H3;23C;25cm;500L ♀ ➤ ◑ ✄   Cichlasoma sp. (Tamasopo) pH7;H3;24C;15cm;300L ♀ ➤ ◑ ✄ 🖬

Cichlasoma cyanoguttatum pH7;H3;23C;25cm;500L ♀ ➤ ◐ ✕ 🖼 ⊡

Cichlasoma regani pH7;H3;25C;25cm;500L ♀ ➤ ◐ ✕ 🖼 ⊡

Cichlasoma meeki pH7;H5;25C;12cm;100L ♀ ➤ ◐ ✕ 🖼 ⊡

Cichlasoma sp. pH7;25C;200L ⌇ ➤ ◐ ✕ 🖼 ⊡

Cichlasoma istlanum pH7;H3;25C;36cm;800L ⌇ ➤ ◐ ✕ 🖼 ⊡

Cichlasoma labridens pH7;H3;25C;25cm;600L ♀ ➤ ◐ ✕ 🖼 ⊡

Cichlasoma regani pH7;H3;25C;25cm;500L ♀ ➤ ◐ ✕ 🖼 ⊡

Cichlasoma labridens pH7;H3;25C;25cm;600L ♀ ➤ ◐ ✕ 🖼 ⊡

Cichlasoma coryphaenoides pH5;H1;28C;25cm;400L 〰 ➤ ◐ ♥ 🖼 ⊟

Cichlasoma temporalis pH6;H1;24C;22cm;400L ♀ ➤ ◐ ♥ 🖼 ⊟

Cichlasoma temporalis pH6;H1;24C;22cm;400L ♀ ➤ ◐ ♥ 🖼 ⊟

Cichlasoma temporalis pH6;H1;24C;22cm;400L ♀ ➤ ◐ ♥ 🖼 ⊟

Cichlasoma festae pH7;H10;25C;50cm;800L ♀ ➤ ◐ ✕ 🖼 ⊟

Cichlasoma festae pH7;H10;25C;50cm;800L ♀ ➤ ◐ ✕ 🖼 ⊟

Cichlasoma umbriferum pH7;H3;28C;50cm;1000L 〰 ➤ ◐ ✕ 🖼 ⊟

Cichlasoma umbriferum pH7;H3;28C;50cm;1000L 〰 ➤ ◐ ✕ 🖼 ⊟

Cichlasoma diquis pH7;H3;26C;20cm;400L ♀ 🐟 ◑ ✂ 🖼 🖽   *Cichlasoma diquis* pH7;H3;26C;20cm;400L ♀ 🐟 ◑ ✂ 🖼 🖽

Cichlasoma dovii pH7;H3;24C;50cm;800L ↘ 🐟 ◑ ✂ 🖼 🖽   *Cichlasoma dovii* pH7;H3;24C;50cm;800L ↘ 🐟 ◑ ✂ 🖼 🖽

Cichlasoma dovii pH7;H3;24C;50cm;800L ↘ 🐟 ◑ ✂ 🖼 🖽   *Cichlasoma dovii* pH7;H3;24C;50cm;800L ↘ 🐟 ◑ ✂ 🖼 🖽

Cichlasoma ellioti pH7;H3;24C;15cm;200L ↘ 🐟 ◑ ✂ 🖼 🖽   *Cichlasoma ellioti* pH7;H3;24C;15cm;200L ↘ 🐟 ◑ ✂ 🖼 🖽

191

Cichlasoma fenestratum pH7;H3;24C;30cm;600L 🌱 🐟 ◑ ✕ 🖼 🖃

Cichlasoma fenestratum pH7;H3;24C;30cm;600L 🌱 🐟 ◑ ✕ 🖼 🖃

Cichlasoma festae pH7;H10;25C;50cm;800L 🌱 🐟 ◑ ✕ 🖼 🖃

Cichlasoma festae pH7;H10;25C;50cm;800L 🌱 🐟 ◑ ✕ 🖼 🖃

Cichlasoma friedrichsthalii pH7;H3;25C;30cm;500L 🌱 🐟 ◑ ✕ 🖼 🖃

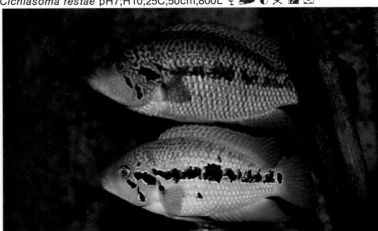

Cichlasoma friedrichsthalii pH7;H3;25C;30cm;500L 🌱 🐟 ◑ ✕ 🖼 🖃

Cichlasoma godmanni pH7;H3;25C;20cm;400L 🌱 🐟 ◑ ✕ 🖼 🖃

Cichlasoma godmanni pH7;H3;25C;20cm;400L 🌱 🐟 ◑ ✕ 🖼 🖃

Cichlasoma grammodes pH7;H15;25C;20cm;300L 〰 🐟 ◑ ✂ 🖼 ▭   *Cichlasoma grammodes* pH7;H15;25C;20cm;300L 〰 🐟 ◑ ✂ 🖼 ▭

Cichlasoma guttulatum pH7;H5;25C;25cm;400L ⚲ 🐟 ◑ ✂ 🖼 ▭   *Cichlasoma guttulatum* pH7;H5;25C;25cm;400L ⚲ 🐟 ◑ ✂ 🖼 ▭

Cichlasoma guttulatum pH7;H5;25C;25cm;400L ⚲ 🐟 ◑ ✂ 🖼 ▭   *Cichlasoma guttulatum* pH7;H5;25C;25cm;400L ⚲ 🐟 ◑ ✂ 🖼 ▭

Cichlasoma hartwegi pH7;H5;25C;25cm;400L ⚲ 🐟 ◑ ♥ 🖼 ▭   *Cichlasoma hartwegi* pH7;H5;25C;25cm;400L ⚲ 🐟 ◑ ♥ 🖼 ▭

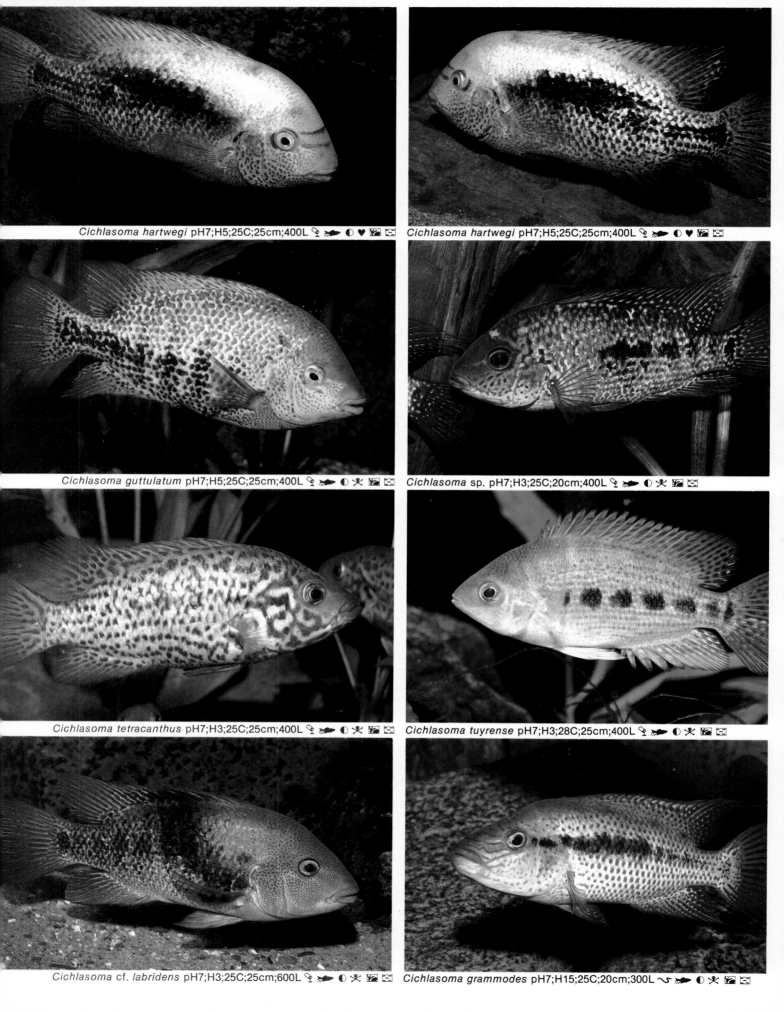

Cichlasoma hartwegi pH7;H5;25C;25cm;400L

Cichlasoma hartwegi pH7;H5;25C;25cm;400L

Cichlasoma guttulatum pH7;H5;25C;25cm;400L

Cichlasoma sp. pH7;H3;25C;20cm;400L

Cichlasoma tetracanthus pH7;H3;25C;25cm;400L

Cichlasoma tuyrense pH7;H3;28C;25cm;400L

Cichlasoma cf. labridens pH7;H3;25C;25cm;600L

Cichlasoma grammodes pH7;H15;25C;20cm;300L

Cichlasoma helleri pH7;H3;25C;15cm;200L ∿ ➤ ◑ ✖ ▨ ⊟

Cichlasoma helleri pH7;H3;25C;15cm;200L ∿ ➤ ◑ ✖ ▨ ⊟

Cichlasoma helleri pH7;H3;25C;15cm;200L ∿ ➤ ◑ ✖ ▨ ⊟

Cichlasoma heterospilus pH7;H3;25C;25cm;500L ♀ ➤ ◑ ✖ ▨ ⊟

Cichlasoma intermedium pH7;H3;25C;25cm;400L ♀ ➤ ◑ ✖ ▨ ⊟

Cichlasoma intermedium pH7;H3;25C;25cm;400L ♀ ➤ ◑ ✖ ▨ ⊟

Cichlasoma intermedium pH7;H3;25C;25cm;400L ♀ ➤ ◑ ✖ ▨ ⊟   Cichlasoma intermedium pH7;H3;25C;25cm;400L ♀ ➤ ◑ ✖ ▨ ⊟

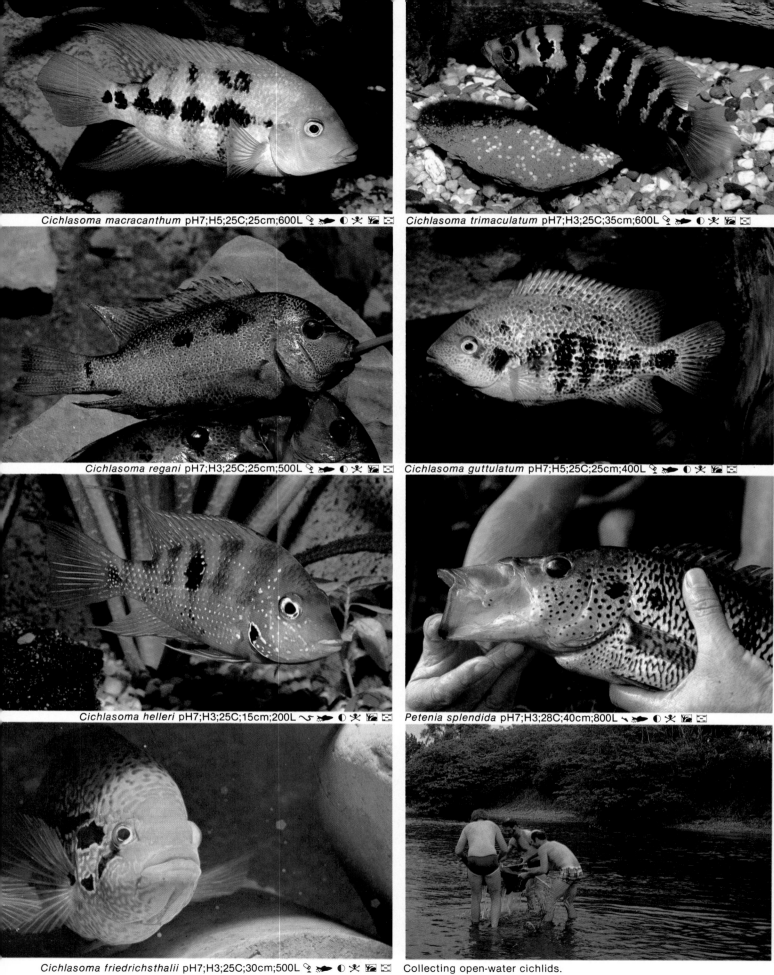

Cichlasoma macracanthum pH7;H5;25C;25cm;600L

Cichlasoma trimaculatum pH7;H3;25C;35cm;600L

Cichlasoma regani pH7;H3;25C;25cm;500L

Cichlasoma guttulatum pH7;H5;25C;25cm;400L

Cichlasoma helleri pH7;H3;25C;15cm;200L

Petenia splendida pH7;H3;28C;40cm;800L

Cichlasoma friedrichsthalii pH7;H3;25C;30cm;500L

Collecting open-water cichlids.

Cichlasoma irregulare pH7;H3;25C;26cm;500L 🐟🌙✂️📷🎞

Cichlasoma istlanum pH7;H3;25C;36cm;800L 🐟🌙✂️📷🎞

Cichlasoma labiatum pH7;H3;25C;25cm;600L 🐟🌙✂️📷🎞

Cichlasoma labiatum pH7;H3;25C;25cm;600L 🐟🌙✂️📷🎞

Cichlasoma labiatum pH7;H3;25C;25cm;600L 🐟🌙✂️📷🎞

Cichlasoma labiatum pH7;H3;25C;25cm;600L 🐟🌙✂️📷🎞

Cichlasoma labridens pH7;H3;25C;25cm;600L 🐟🌙✂️📷🎞

Cichlasoma labridens pH7;H3;25C;25cm;600L 🐟🌙✂️📷🎞

Cichlasoma lentiginosum pH7;H15;25C;26cm;600L ♀ ➤ ◑ ✕ 🖼 ⬚   *Cichlasoma lentiginosum* pH7;H15;25C;26cm;600L ♀ ➤ ◑ ✕ 🖼 ⬚

Cichlasoma lentiginosum pH7;H15;25C;26cm;600L ♀ ➤ ◑ ✕ 🖼 ⬚   *Cichlasoma longimanus* pH7;H5;27C;14cm;400L ♀ ➤ ◑ ✕ 🖼 ⬚

Cichlasoma longimanus pH7;H5;27C;14cm;400L ♀ ➤ ◑ ✕ 🖼 ⬚   *Cichlasoma longimanus* pH7;H5;27C;14cm;400L ♀ ➤ ◑ ✕ 🖼 ⬚

Cichlasoma longimanus pH7;H5;27C;14cm;400L ♀ ➤ ◑ ✕ 🖼 ⬚   *Cichlasoma lyonsi* pH7;H5;25C;25cm;600L ♀ ➤ ◑ ✕ 🖼 ⬚

Cichlasoma macracanthum pH7;H5;25C;25cm;600L 🌿🐟◑✂🖼️🎞️

Cichlasoma macracanthum pH7;H5;25C;25cm;600L 🌿🐟◑✂🖼️🎞️

Cichlasoma macracanthum pH7;H5;25C;25cm;600L 🌿🐟◑✂🖼️🎞️

Cichlasoma maculicauda pH7;H5;25C;25cm;600L 🌿🐟◑✂🖼️🎞️

Cichlasoma maculicauda pH7;H5;25C;25cm;600L 🌿🐟◑✂🖼️🎞️

Cichlasoma maculicauda pH7;H5;25C;25cm;600L 🌿🐟◑✂🖼️🎞️

Cichlasoma managuense pH7;H5;25C;50cm;800L 🔨🐟◑✂🖼️🎞️

Cichlasoma managuense pH7;H5;25C;50cm;800L 🔨🐟◑✂🖼️🎞️

Cichlasoma meeki pH7;H5;25C;12cm;100L ♀ 🐟 🌓 ✗ 🖼 🖵
Cichlasoma meeki pH7;H5;25C;12cm;100L ♀ 🐟 🌓 ✗ 🖼 🖵

Cichlasoma melanurus pH7;H5;25C;25cm;500L ♀ 🐟 🌓 ✗ 🖼 🖵
Cichlasoma melanurus pH7;H5;25C;25cm;500L ♀ 🐟 🌓 ✗ 🖼 🖵

Cichlasoma microphthalmus pH7;H5;25C;25cm;500L ♀ 🐟 🌓 ✗
Cichlasoma minckleyi pH8;H50;24C;17cm;300L ♀ 🐟 🌓 ✗ 🖼 🖵

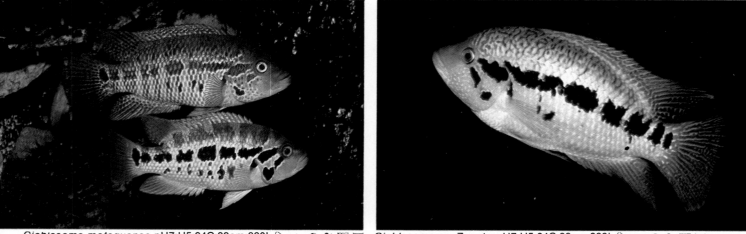

Cichlasoma motaguense pH7;H5;24C;30cm;600L ♀ 🐟 🌓 ✗ 🖼 🖵
Cichlasoma sp. Zapote pH7;H5;24C;30cm;600L ♀ 🐟 🌓 ✗ 🖼 🖵

Cichlasoma carpinte pH7;H3;25C;20cm;500L ♀ ➤ ◑ ✕ 🎞 ⊟

Cichlasoma carpinte pH7;H3;25C;20cm;500L ♀ ➤ ◑ ✕ 🎞 ⊟

Cichlasoma carpinte pH7;H3;25C;20cm;500L ♀ ➤ ◑ ✕ 🎞 ⊟

Cichlasoma carpinte pH7;H3;25C;20cm;500L ♀ ➤ ◑ ✕ 🎞 ⊟

Cichlasoma minckleyi pH8;H50;24C;17cm;300L ♀ ➤ ◑ ✕ 🎞 ⊟

Cichlasoma minckleyi pH8;H50;24C;17cm;300L ♀ ➤ ◑ ✕ 🎞

Cichlasoma octofasciatum pH7;H5;24C;25cm;400L ♀ ➤ ◑ ✕ 🎞 ⊟

Cichlasoma octofasciatum pH7;H5;24C;25cm;400L ♀ ➤ ◑ ✕ 🎞 ⊟

Cichlasoma nigrofasciatum pH7;H5;24C;12cm;100L

Cichlasoma nigrofasciatum pH7;H5;24C;12cm;100L

Cichlasoma nigrofasciatum pH7;H5;24C;12cm;100L

Cichlasoma centrarchus pH7;H3;25C;15cm;300L

Cichlasoma octofasciatum pH7;H5;24C;25cm;400L

Cichlasoma octofasciatum pH7;H5;24C;25cm;400L

Cichlasoma longimanus pH7;H5;27C;14cm;400L

Cichlasoma longimanus pH7;H5;27C;14cm;400L

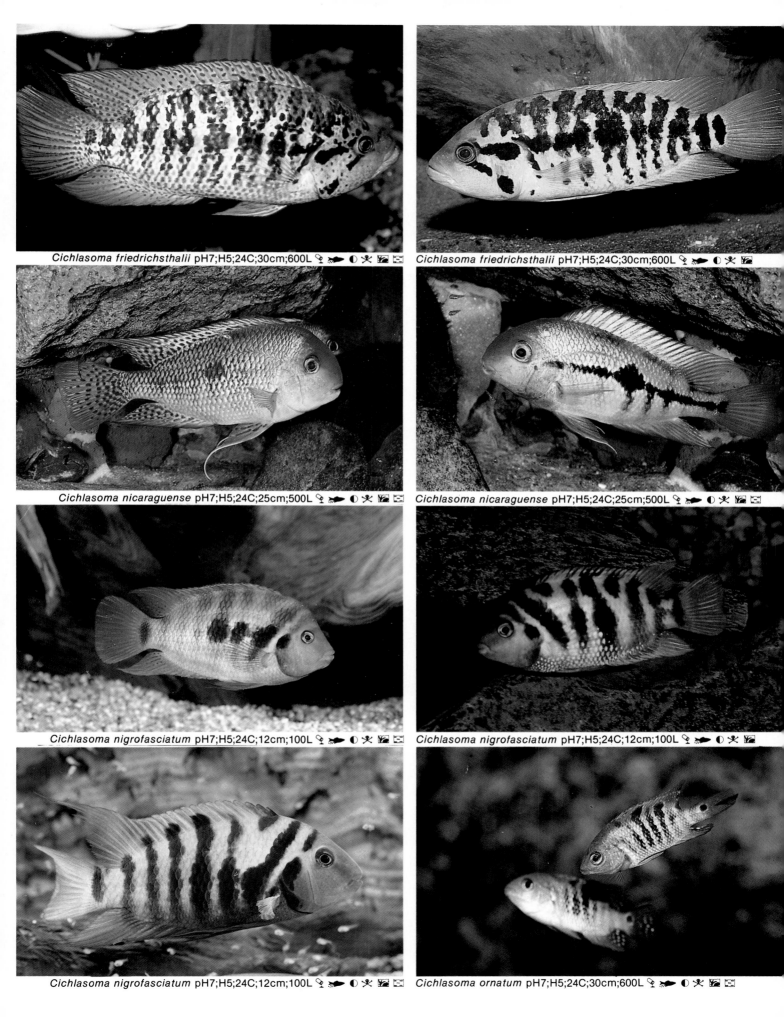

Cichlasoma friedrichsthalii pH7;H5;24C;30cm;600L ♀ ➥ ◑ ✂ 🖼 ▱

Cichlasoma friedrichsthalii pH7;H5;24C;30cm;600L ♀ ➥ ◑ ✂ 🖼 ▱

Cichlasoma nicaraguense pH7;H5;24C;25cm;500L ♀ ➥ ◑ ✂ 🖼 ▱

Cichlasoma nicaraguense pH7;H5;24C;25cm;500L ♀ ➥ ◑ ✂ 🖼 ▱

Cichlasoma nigrofasciatum pH7;H5;24C;12cm;100L ♀ ➥ ◑ ✂ 🖼 ▱

Cichlasoma nigrofasciatum pH7;H5;24C;12cm;100L ♀ ➥ ◑ ✂ 🖼

Cichlasoma nigrofasciatum pH7;H5;24C;12cm;100L ♀ ➥ ◑ ✂ 🖼 ▱

Cichlasoma ornatum pH7;H5;24C;30cm;600L ♀ ➥ ◑ ✂ 🖼 ▱

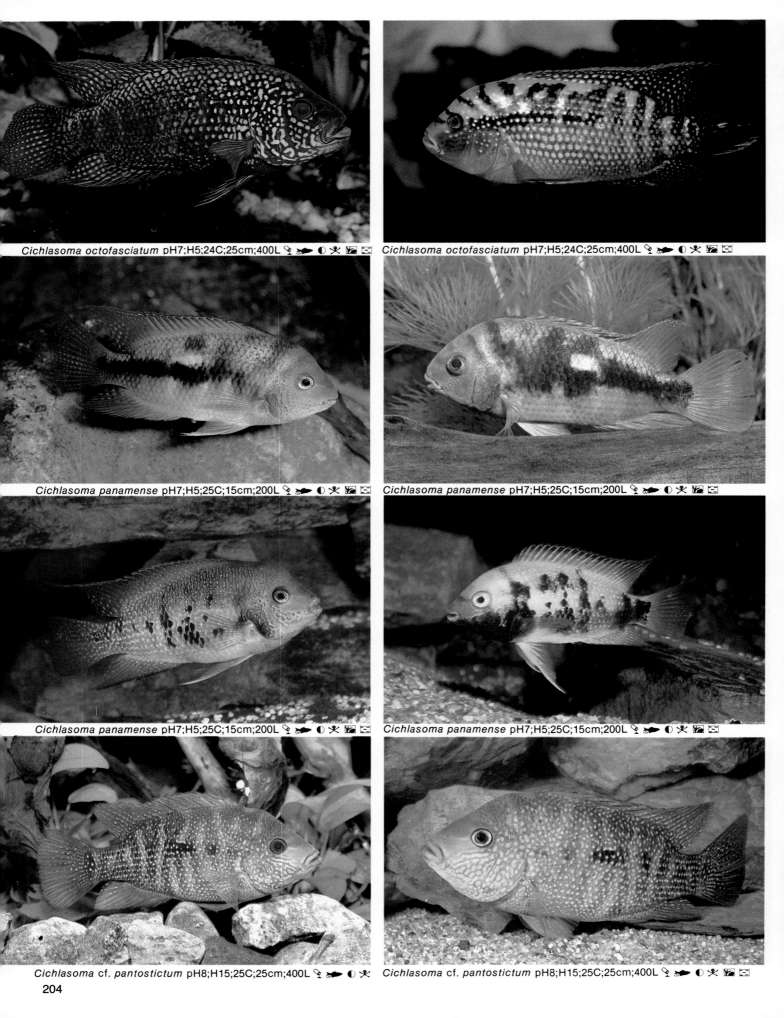

Cichlasoma octofasciatum pH7;H5;24C;25cm;400L ♀ ➤ ◑ ✕ ▥ ▣

Cichlasoma octofasciatum pH7;H5;24C;25cm;400L ♀ ➤ ◑ ✕ ▥ ▣

Cichlasoma panamense pH7;H5;25C;15cm;200L ♀ ➤ ◑ ✕ ▥ ▣

Cichlasoma panamense pH7;H5;25C;15cm;200L ♀ ➤ ◑ ✕ ▥ ▣

Cichlasoma panamense pH7;H5;25C;15cm;200L ♀ ➤ ◑ ✕ ▥ ▣

Cichlasoma panamense pH7;H5;25C;15cm;200L ♀ ➤ ◑ ✕ ▥ ▣

Cichlasoma cf. *pantostictum* pH8;H15;25C;25cm;400L ♀ ➤ ◑ ✕

Cichlasoma cf. *pantostictum* pH8;H15;25C;25cm;400L ♀ ➤ ◑ ✕ ▥ ▣

Cichlasoma pasionis pH7;H5;25C;15cm;200L ∿ 🐟 ◐ ✕ 🖼 ▭ Cichlasoma pasionis pH7;H5;25C;15cm;200L ∿ 🐟 ◐ ✕ 🖼 ▭

Cichlasoma pearsei pH7;H5;25C;25cm;600L ♀ 🐟 ◐ ✕ 🖼 ▭ Cichlasoma cf. bocourti pH7;H3;25C;25cm;600L ♀ 🐟 ◐ ✕ 🖼 ▭

Cichlasoma regani pH7;H3;25C;25cm;500L ♀ 🐟 ◐ ✕ 🖼 ▭ Cichlasoma robertsoni pH7;H3;25C;20cm;400L ♀ 🐟 ◐ ✕ 🖼 ▭

Cichlasoma robertsoni pH7;H3;25C;20cm;400L ♀ 🐟 ◐ ✕ 🖼 ▭ Cichlasoma robertsoni pH7;H3;25C;20cm;400L ♀ 🐟 ◐ ✕ 🖼 ▭

Collecting cichlids near dense vegetation.

Cichlasoma cf. *aureum* pH7;H3;25C;14cm;300L 〰 🐟 ◑ ✕ 🖼 ⊟

Cichlasoma robertsoni pH7;H3;25C;20cm;400L ♀ 🐟 ◑ ✕ 🖼 ⊟

Cichlasoma bifasciatum pH7;H3;25C;30cm;800L ♀ 🐟 ◑ ✕ 🖼 ⊟

Cichlasoma octofasciatum pH7;H5;24C;25cm;400L 🐟 ◑ ✕ 🖼

Cichlasoma aff. *pearsei* pH7;H5;25C;25cm;600L ♀ 🐟 ◑ ✕ 🖼 ⊟

Cichlasoma rostratum pH7;H3;25C;24cm;500L ♀ 🐟 ◑ ✂ 🖼 ⊡

Cichlasoma rostratum pH7;H3;25C;24cm;500L ♀ 🐟 ◑ ✂ 🖼 ⊡

Cichlasoma rostratum pH7;H3;25C;24cm;500L ♀ 🐟 ◑ ✂ 🖼 ⊡

Cichlasoma sajica pH7;H10;25C;12cm;100L ♀ 🐟 ◑ ✂ 🖼 ⊡

Cichlasoma sajica pH7;H10;25C;12cm;100L ♀ 🐟 ◑ ✂ 🖼 ⊡

Cichlasoma sajica pH7;H10;25C;12cm;100L ♀ 🐟 ◑ ✂ 🖼 ⊡

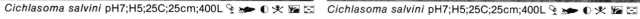

Cichlasoma salvini pH7;H5;25C;25cm;400L ♀ 🐟 ◑ ✂ 🖼 ⊡

Cichlasoma salvini pH7;H5;25C;25cm;400L ♀ 🐟 ◑ ✂ 🖼 ⊡

Cichlasoma salvini pH7;H5;25C;25cm;400L ♀ 🐟 ◑ ✕ 🖼 ⊟

Cichlasoma motaguense pH7;H5;24C;30cm;600L ♀ 🐟 ◑ ✕ 🖼 ⊟

Cichlasoma motaguense pH7;H5;24C;30cm;600L ♀ 🐟 ◑ ✕ 🖼 ⊟

Cichlasoma friedrichsthalii pH7;H3;25C;30cm;500L ♀ 🐟 ◑ ✕ 🖼 ⊟

Cichlasoma managuense pH7;H5;25C;50cm;800L ↘ 🐟 ◑ ✕ 🖼 ⊟

Cichlasoma managuense pH7;H5;25C;50cm;800L ↘ 🐟 ◑ ✕ 🖼 ⊟

Cichlasoma dovii pH7;H3;24C;50cm;800L ↘ 🐟 ◑ ✕ 🖼 ⊟

Cichlasoma dovii pH7;H3;24C;50cm;800L ↘ 🐟 ◑ ✕ 🖼 ⊟

208

Cichlasoma septemfasciatum pH7;H5;25C;12cm;100L ♀ 🐟 ◑ ✂ 🖼 ⊟

Cichlasoma septemfasciatum pH7;H5;25C;12cm;100L ♀ 🐟 ◑ ✂

Cichlasoma septemfasciatum pH7;H5;25C;12cm;100L ♀ 🐟 ◑ ✂ 🖼 ⊟

Cichlasoma septemfasciatum pH7;H5;25C;12cm;100L ♀ 🐟 ◑ ✂

Cichlasoma septemfasciatum pH7;H5;25C;12cm;100L ♀ 🐟 ◑ ✂ 🖼 ⊟

Cichlasoma septemfasciatum pH7;H5;25C;12cm;100L ♀ 🐟 ◑ ✂

Cichlasoma sieboldii pH7;H3;25C;15cm;200L ♀ 🐟 ◑ ✂ 🖼 ⊟

Cichlasoma sieboldii pH7;H3;25C;15cm;200L ♀ 🐟 ◑ ✂ 🖼 ⊟

Cichlasoma socolofi pH7;H3;25C;12cm;100L ∿ ➤ ◑ ✕ 🖼 ⊟

Cichlasoma spilurus pH7;H3;25C;12cm;100L ♀ ➤ ◑ ✕ 🖼 ⊟

Cichlasoma spilurus pH7;H3;25C;12cm;100L ♀ ➤ ◑ ✕ 🖼 ⊟

Cichlasoma spilurus pH7;H3;25C;12cm;100L ♀ ➤ ◑ ✕ 🖼 ⊟

Cichlasoma spilurus pH7;H3;25C;12cm;100L ♀ ➤ ◑ ✕ 🖼 ⊟

Cichlasoma spilurus pH7;H3;25C;12cm;100L ♀ ➤ ◑ ✕ 🖼 ⊟

Cichlasoma spilurus pH7;H3;25C;12cm;100L ♀ ➤ ◑ ✕ 🖼 ⊟

Cichlasoma steindachneri pH7;H3;25C;20cm;300L ∿ ➤ ◑ ✕ 🖼 ⊟

Cichlasoma psittacus pH6;H1;26C;30cm;50L ～ ➤ ◑ ♥ 🎞 ⊟   *Cichlasoma sajica* pH7;H10;25C;12cm;100L ⚲ ➤ ◑ ✄ 🎞 ⊟

Cichlasoma septemfasciatum pH7;H5;25C;12cm;100L ⚲ ➤ ◑ ✄ 🎞 ⊟   *Cichlasoma septemfasciatum* pH7;H5;25C;12cm;100L ⚲ ➤ ◑ ✄ 🎞

Cichlasoma septemfasciatum pH7;H5;25C;12cm;100L ⚲ ➤ ◑ ✄ 🎞 ⊟   *Cichlasoma spilurus* pH7;H3;25C;10cm;100L ⚲ ➤ ◑ ✄ 🎞 ⊟

Cichlasoma spilurus pH7;H3;25C;10cm;100L ⚲ ➤ ◑ ✄ 🎞 ⊟   *Cichlasoma spilurus* pH7;H3;25C;10cm;100L ⚲ ➤ ◑ ✄ 🎞 ⊟

211

Cichlasoma synspilus pH7;H3;25C;30cm;500L ♀ ⤚ ◑ ✕ 🖾 ⊡　*Cichlasoma synspilus* pH7;H3;25C;30cm;500L ♀ ⤚ ◑ ✕ 🖾 ⊡

Cichlasoma synspilus pH7;H3;25C;30cm;500L ♀ ⤚ ◑ ✕ 🖾 ⊡　*Cichlasoma synspilus* pH7;H3;25C;30cm;500L ♀ ⤚ ◑ ✕ 🖾 ⊡

Cichlasoma tetracanthus pH7;H3;25C;25cm;400L ♀ ⤚ ◑ ✕ 🖾 ⊡　*Cichlasoma tetracanthus* pH7;H3;25C;25cm;400L ♀ ⤚ ◑ ✕ 🖾 ⊡

Cichlasoma trimaculatus pH7;H3;25C;35cm;600L ♀ ⤚ ◑ ✕ 🖾 ⊡　*Cichlasoma trimaculatus* pH7;H3;25C;35cm;600L ♀ ⤚ ◑ ✕ 🖾 ⊡

Cichlasoma synspilus pH7;H3;25C;30cm;500L ♀ 🐟 ◑ ✕ 🖼 ▤

Cichlasoma synspilus pH7;H3;25C;30cm;500L ♀ 🐟 ◑ ✕ 🖼 ▤

Cichlasoma maculicauda pH7;H5;25C;25cm;600L ♀ 🐟 ◑ ✕ 🖼 ▤

Cichlasoma nicaraguense pH7;H5;24C;25cm;500L ♀ 🐟 ◑ ✕ 🖼

Cichlasoma nicaraguense pH7;H5;24C;25cm;500L ♀ 🐟 ◑ ✕ 🖼 ▤

Cichlasoma nicaraguense pH7;H5;24C;25cm;500L ♀ 🐟 ◑ ✕ 🖼

Cichlasoma spilurus pH7;H3;25C;12cm;100L ♀ 🐟 ◑ ✕ 🖼 ▤

Cichlasoma spilurus pH7;H3;25C;12cm;100L ♀ 🐟 ◑ ✕ 🖼 ▤

213

Cichlasoma tuba pH7;H3;25C;25cm;400L ♀ ➤ ◑ ✕ 🎞 ⊟

Cichlasoma tuba pH7;H3;25C;25cm;400L ♀ ➤ ◑ ✕ 🎞 ⊟

Cichlasoma tuba pH7;H3;25C;25cm;400L ♀ ➤ ◑ ✕ 🎞 ⊟

Cichlasoma tuba pH7;H3;25C;25cm;400L ♀ ➤ ◑ ✕ 🎞 ⊟

Cichlasoma tuyrense pH7;H3;28C;25cm;400L ♀ ➤ ◑ ✕ 🎞 ⊟

Cichlasoma tuyrense pH7;H3;28C;25cm;400L ♀ ➤ ◑ ✕ 🎞 ⊟

Cichlasoma umbriferum pH7;H3;28C;50cm;1000L ↘ ➤ ◑ ✕ 🎞 ⊟

Cichlasoma umbriferum pH7;H3;28C;50cm;1000L ↘ ➤ ◑ ✕ 🎞 ⊟

Cichlasoma urophthalmus pH7;H3;28C;30cm;600L ♀ ⊱ ◑ ✻ ▦ ⊟   *Cichlasoma urophthalmus* pH7;H3;28C;30cm;600L ♀ ⊱ ◑ ✻ ▦ ⊟

Cichlasoma urophthalmus pH7;H3;28C;30cm;600L ♀ ⊱ ◑ ✻ ▦ ⊟   *Cichlasoma zaliosum* pH7;H3;28C;20cm;400L ♀ ⊱ ◑ ✻ ▦ ⊟

Neetroplus nematopus pH7;H10;25C;15cm;200L ♀ ⊱ ◑ ✻ ▦ ⊟   *Neetroplus nematopus* pH7;H10;25C;15cm;200L ♀ ⊱ ◑ ✻ ▦ ⊟

Petenia splendida pH7;H3;28C;40cm;800L ⬈ ⊱ ◑ ✻ ▦ ⊟   *Petenia splendida* pH7;H3;28C;40cm;800L ⬈ ⊱ ◑ ✻ ▦ ⊟

Cichlasoma spilurus pH7;H3;25C;12cm;100L ♀ ⬳ ◑ ✕ ▣ ⊟

Cichlasoma umbriferum pH7;H3;28C;50cm;1000L ⬳ ◑ ✕ ▣ ⊟

Cichlasoma sp. Sambu pH7;H3;28C;50cm;1000L ⬳ ◑ ✕ ▣ ⊟

Cichlasoma panamense pH7;H5;25C;15cm;200L ♀ ⬳ ◑ ✕ ▣ ⊟

Herotilapia multispinosa pH7;H10;24C;10cm;100L ♀ ⬳ ◑ ♥ ▣ ⊟

Cichlasoma tetracanthus pH7;H3;25C;25cm;400L ♀ ⬳ ◑ ✕ ▣ ⊟

Herotilapia multispinosa pH7;H10;24C;10cm;100L ♀ ⬳ ◑ ♥ ▣ ⊟

Herotilapia multispinosa pH7;H10;24C;10cm;100L ♀ ⬳ ◑ ♥ ▣ ⊟

Petenia splendida pH7;H3;28C;40cm;800L ⤵ 🐟 ◑ ✗ 🖼 ⊟

Cichlasoma kraussii pH6;H1;24C;25cm;500L ⤳ 🐟 ◑ ✗ 🖼 ⊟

Petenia splendida pH7;H3;28C;40cm;800L ⤵ 🐟 ◑ ✗ 🖼 ⊟

Cichlasoma umbriferum pH7;H3;28C;50cm;1000L ⤵ 🐟 ◑ ✗ 🖼 ⊟

Neetroplus nematopus pH7;H10;25C;15cm;200L ⚲ 🐟 ◑ ✗ 🖼 ⊟

Herotilapia multispinosa pH7;H10;24C;10cm;100L ⚲ 🐟 ◑ ♥ 🖼 ⊟

Herotilapia multispinosa pH7;H10;24C;10cm;100L ⚲ 🐟 ◑ ♥ 🖼 ⊟

Herotilapia multispinosa pH7;H10;24C;10cm;100L ⚲ 🐟 ◑ ♥ 🖼 ⊟

217

The following checklist and that of the synonyms were compiled using the *Cichlid-Catalogue* of Ufermann, Allgayer, and Geerts.

Aequidens coeruleopunctatus (KNER & STEINDACHNER, 1863)

Cichlasoma (Thorichthys) affinis (GUENTHER, 1862)

Cichlasoma (Amphilophus) alfari MEEK, 1907

Cichlasoma (Amphilophus) altifrons (KNER & STEINDACHNER, 1863)

Cichlasoma (Parapetenia) atromaculatum (REGAN, 1912)

Cichlasoma (Thorichthys) aureum (GUENTHER, 1862)

Cichlasoma (Parapetenia) bartoni (BEAN, 1892)

Cichlasoma (Parapetenia) beani (JORDAN, 1888)

Cichlasoma (Theraps) bifasciatum (STEINDACHNER, 1864)

Cichlasoma (Herichthys) bocourti (VAILLANT & PELLEGRIN, 1902)

Cichlasoma (Theraps) breidhori (WERNER & STAWIKOWSKI, 1987)

Cichlasoma (Paraneetroplus) bulleri (REGAN, 1905)

Cichlasoma (Thorichthys) callolepis (REGAN, 1904)

Cichlasoma (Amphilophus) calobrense MEEK & HILDEBRAND, 1913

Cichlasoma (Herichthys) carpinte (JORDAN & SNYDER, 1899)

Cichlasoma (Archocentrus) centrarchus (GILL & BRANSFORD, 1877)

Cichlasoma (Amphilophus) citrinellum (GUENTHER, 1864)

Cichlasoma (Theraps) coeruleus (STAWIKOWSKI & WERNER, 1987)

Cichlasoma (Herichthys) cyanoguttatum (BAIRD & GIRARD, 1854)

Cichlasoma (Amphilophus) diquis BUSSING, 1974

Cichlasoma (Parapetenia) dovii (GUENTHER, 1864)

Cichlasoma (Thorichthys) ellioti (MEEK, 1904)

Cichlasoma (Theraps) fenestratum (GUENTHER, 1860)

Cichlasoma (Parapetenia) festae (BOULENGER, 1899)

Cichlasoma (Parapetenia) friedrichsthalii (HECKEL, 1840)

Cichlasoma (Theraps) gibbiceps (STEINDACHNER, 1864)

Cichlasoma (Theraps) godmanni (GUENTHER, 1862)

Cichlasoma (Parapetenia) grammodes TAYLOR & MILLER, 1980

Cichlasoma (Theraps) guttulatum (GUENTHER, 1864)

Cichlasoma (Parapetenia) haitiensis TEE-VAN, 1935

Cichlasoma (Theraps) hartwegi TAYLOR & MILLER, 1980

Cichlasoma (Thorichthys) helleri (STEINDACHNER, 1864)

Cichlasoma (Theraps) heterospilus HUBBS, 1936

Cichlasoma (Parapetenia) hogaboomorum CARR & GIOVANNOLI, 1950

Cichlasoma (Theraps) intermedium (GUENTHER, 1862)

Cichlasoma (Theraps) irregulare (GUENTHER, 1862)

Cichlasoma (Parapetenia) istlanum (JORDAN & SNYDER, 1899)

Cichlasoma (Amphilophus) labiatum (GUENTHER, 1864)

Cichlasoma (Parapetenia) labridens (PELLEGRIN, 1903)

Cichlasoma (Theraps) lentiginosum (STEINDACHNER, 1864)

Cichlasoma (Parapetenia) leonhardschultzei AHL, 1935

Cichlasoma (Amphilophus) longimanus (GUENTHER, 1869)

Cichlasoma (Amphilophus) lyonsi GOSSE, 1966

Cichlasoma (Amphilophus) macracanthum (GUENTHER, 1864)

Cichlasoma (Theraps) maculicauda (REGAN, 1905)

Cichlasoma (Parapetenia) managuense (GUENTHER, 1869)

Cichlasoma (Amphilophus) margaritiferum (GUENTHER, 1862)

Cichlasoma (Thorichthys) meeki (BRIND, 1918)

Cichlasoma (Theraps) melanurus (GUENTHER, 1862)

Cichlasoma (Theraps) microphthalmus (GUENTHER, 1862)

Cichlasoma (Parapetenia) minckleyi KORNFIELD & TAYLOR, 1983

Cichlasoma (Parapetenia) motaguense (GUENTHER, 1866)

Cichlasoma (Theraps) nebuliferum (GUENTHER, 1860)

Cichlasoma (Theraps) nicaraguense (GUENTHER, 1864)

Cichlasoma (Archocentrus) nigrofasciatum (GUENTHER, 1869)

Cichlasoma (Parapetenia) octofasciatum (REGAN, 1903)

Cichlasoma (Parapetenia) ornatum (REGAN, 1905)

Cichlasoma (Theraps) panamense (MEEK & HILDEBRAND, 1913)

Cichlasoma (Parapetenia) pantostictum TAYLOR & MILLER, 1983

Cichlasoma (Thorichthys) pasionis RIVAS, 1962

Cichlasoma (Herichthys)

pearsei (HUBBS, 1936)

Cichlasoma (Parapetenia) ramsdeni FOWLER, 1938

Cichlasoma (Theraps) regani MILLER, 1974

Cichlasoma (Amphilophus) rhytisma LOPEZ, 1983

Cichlasoma (Amphilophus) robertsoni (REGAN, 1905)

Cichlasoma (Amphilophus) rostratum (GILL & BRANSFORD, 1877)

Cichlasoma (Archocentrus) sajica BUSSING, 1974

Cichlasoma (Parapetenia) salvini (GUENTHER, 1862)

Cichlasoma (Archocentrus) septemfasciatum (REGAN, 1908)

Cichlasoma (Theraps) sieboldii (KNER & STEINDACHNER, 1863)

Cichlasoma (Thorichthys) socolofi MILLER & TAYLOR, 1984

Cichlasoma (Archocentrus) spilurus (GUENTHER, 1862)

Cichlasoma (Archocentrus) spinosissimus (VAILLANT & PELLEGRIN, 1902)

Cichlasoma (Theraps) steindachneri JORDAN & SNYDER, 1899

Cichlasoma (Theraps) synspilus HUBBS, 1935

Cichlasoma (Parapetenia) tetracanthus (VALENCIENNES, 1831)

Cichlasoma (Parapetenia) trimaculatus (GUENTHER, 1869)

Cichlasoma (Tomocichla) tuba MEEK, 1912

Cichlasoma (Amphilophus) tuyrense MEEK & HILDEBRAND, 1913

Cichlasoma (Parapetenia) umbriferum MEEK & HILDEBRAND, 1913

Cichlasoma (Parapetenia) urophthalmus (GUENTHER, 1862)

Cichlasoma (Parapetenia) vombergae LADIGES, 1938

Cichlasoma (Amphilophus) zaliosum BARLOW, 1976

Geophagus crassilabris STEINDACHNER, 1877

Herotilapia multispinosa (GUENTHER, 1866)

Neetroplus nematopus GUENTHER, 1866

Petenia splendida GUENTHER, 1862

The following list contains all known synonyms of Central American cichlids that appeared in the scientific literature. The specific (second) names are in alphabetical order since many synonyms were used in combination with several generic names.

Cichlasoma acutum MILLER, 1907 = *Cichlasoma robertsoni*

Acara adspersa GUENTHER, 1862 = *Cichlasoma tetracanthus*

Cichlasoma alfaroi MEEK, 1907 = *Cichlasoma alfari*

Heros angulifer GUENTHER, 1862 = *Cichlasoma intermedium*

Heros balteatus GILL & BRANSFORD, 1877 = *Cichlasoma nicaraguense*

Heros basilaris GILL & BRANSFORD, 1877 = *Cichlasoma citrinellum*

Cichlasoma biocellatum REGAN, 1909 = *Cichlasoma octofasciatum*

Cichlasoma bouchellei FOWLER, 1923 = *Cichlasoma alfari*

Cichlasoma caeruleogula FOWLER, 1935 = *Cichlasoma microphthalmus*

Cichlasoma cajali ALVAREZ & GUTIERREZ, 1952 = *Cichlasoma trimaculatum*

Cichlasoma centrale MEEK, 1906 = *Cichlasoma trimaculatum*

Cichlasoma champotonis HUBBS, 1936 = *Cichlasoma helleri*

Acara cubensis (HECKEL) STEINDACHNER, 1863 = *Cichlasoma tetracanthus*

Cichlasoma cutteri FOWLER, 1932 = *Cichlasoma spilurus*

Heros deppii HECKEL, 1840 = *Cichlasoma sieboldii*

Cichlasoma dorsatum MEEK, 1907 = *Cichlasoma labiatum*

Cichlasoma eigenmanni MEEK, 1902 = *Cichlasoma*

nebuliferum

Heros erythraeus GUENTHER, 1866 = *Cichlasoma labiatum*

Cichlasoma evermanni MEEK, 1904 = *Cichlasoma trimaculatum*

Neetroplus fluviatilis MEEK, 1912 = *Neetroplus nematopus*

Amphilophus froebelii AGASSIZ, 1858 = *Cichlasoma labiatum*

Cichlasoma frontale MEEK, 1909 = *Cichlasoma sieboldii*

Cichlasoma frontosa MEEK, 1909 = *Cichlasoma sieboldii*

Chromis fusco-maculatus GUICHENOT, 1850 = *Cichlasoma tetracanthus*

Cichlasoma gadovii REGAN, 1905 = *Cichlasoma fenestratum*

Cichlasoma geddesi REGAN, 1905 = *Cichlasoma cyanoguttatum*

Cichlasoma globosum MILLER, 1907 = *Cichlasoma maculicauda*

Cichlasoma gordon-smithi FOWLER, 1935 = *Cichlasoma trimaculatum*

Cichlasoma granadense MEEK, 1907 = *Cichlasoma citrinellum*

Cichlasoma guentheri PELLEGRIN, 1904 = *Cichlasoma microphthalmus*

Cichlasoma guija HILDEBRAND, 1934 = *Cichlasoma macracanthum*

Cichlasoma hedricki MEEK, 1904 = *Cichlasoma octofasciatum*

Heros heterodontus VAILLANT & PELLEGRIN, 1902 = *Cichlasoma macracanthum*

Cichlaurus hicklingi FOWLER, 1956 = *Cichlasoma synspilus*

Cichlasoma hyorhynchum HUBBS, 1935 = *Cichlasoma meeki*

Cichlasoma immaculatum PELLEGRIN, 1904 =

Cichlasoma spinosissimus

Cichlasoma laurae REGAN, 1908 = *Cichlasoma cyanoguttatum*

Cichlasoma lethrinus REGAN, 1908 = *Cichlasoma alfari*

Heros lobochilus GUENTHER, 1866 = *Cichlasoma labiatum*

Heros maculipinnis STEINDACHNER, 1864 = *Cichlasoma aureum*

Cichlasoma manana MILLER, 1907 = *Cichlasoma maculicauda*

Heros margaritifer var. STEINDACHNER, 1879 = *Cichlasoma tuyrense*

Heros melanopogon STEINDACHNER, 1864 = *Cichlasoma melanurus*

Heros mento VAILLANT & PELLEGRIN, 1902 = *Cichlasoma istlanum*

Cichlasoma milleri MEEK, 1907 = *Cichlasoma microphthalmus*

Cichlasoma mojarra MEEK, 1904 = *Cichlasoma trimaculatum*

Heros montezuma HECKEL, 1840 = *Cichlasoma sieboldii*

Cichlasoma multifasciatum REGAN, 1905 = *Cichlasoma friedrichsthalii*

Neetroplus nicaraguensis GILL & BRANSFORD, 1877 = *Neetroplus nematopus*

Heros nigricans EIGENMANN, 1902 = *Cichlasoma tetracanthus*

Cichlasoma nigritum MEEK, 1907 = *Cichlasoma labiatum*

Cichlasoma nigritum MEEK, 1907 = *Cichlasoma maculicauda*

Heros oblongus GUENTHER, 1866 = *Cichlasoma microphthalmus*

Vieja panamensis FERNANDEZ-YEPEZ, 1969 = *Cichlasoma maculicauda*

Heros parma GUENTHER, 1862 = *Cichlasoma fenestratum*

Heros pavonaceus GARMAN,

1881 = *Cichlasoma cyanoguttatum*

Cichlasoma popenoei CARR & GIOVANNOLI, 1950 = *Cichlasoma longimanus*

Cichlasoma punctatum MEEK, 1909 = *Cichlasoma sieboldii*

Acara rectangularis STEINDACHNER, 1864 = *Cichlasoma intermedium*

Theraps rheophilus SEEGERS & STAECK, 1985 = *Cichlasoma lentiginosum*

Cichlasoma sexfasciatus REGAN, 1905 = *Cichlasoma fenestratum*

Cichlasoma spilotum MEEK, 1912 = *Cichlasoma nicaraguense*

Cichlasoma teapae EVERMANN & GOLDSBOROUGH, 1902 = *Cichlasoma gibbiceps*

Cichlasoma tenue MEEK, 1906 = *Cichlasoma salvini*

Heros teporatus FOWLER, 1903 = *Cichlasoma cyanoguttatum*

Theraps terrabae JORDAN & EVERMANN, 1927 = *Cichlasoma sieboldii*

Heros triagramma STEINDACHNER, 1864 = *Cichlasoma salvini*

Heros troscheli STEINDACHNER, 1867 = *Cichlasoma urophthalmus*

Herichthys underwoodi REGAN, 1906 = *Cichlasoma sieboldii*

Cichlasoma zonatum MEEK, 1905 = *Cichlasoma guttulatum*

BIBLIOGRAPHY

BARLOW, G.W. (1976) The Midas Cichlid in Nicaragua. In: Investigations of the Ichthyofauna of Nicaraguan Lakes, T.B. Thorson (Ed.), Lincoln, Nebraska: 333-358.

BARLOW, G.W. (1983) Do Gold Midas Cichlid Fish Win Fights Because of Their Colour, or Because They Lack Normal Coloration? Behav. Ecol. Sociobiol. (13): 197-204.

BARLOW, G.W. & MUNSEY, J.W. (1976) The Red Devil-Midas-Arrow Cichlid Species Complex in Nicaragua. In: Investigations of the Ichthyofauna of Nicaraguan Lakes, T.B. Thorson (Ed.), Lincoln, Nebraska: 359-369.

BUSSING, W.A. (1976) Geographic Distribution of the San Juan Ichthyofauna of Central America with Remarks on its Origin and Ecology. In: Investigations of the Ichthyofauna of Nicaraguan Lakes, T.B. Thorson (Ed.), Lincoln, Nebraska

COURTENAY, S.C. & KEENLEYSIDE, M.H.A. (1983) Wriggler-hanging: a response to hypoxia by broodrearing *Herotilapia multispinosa.* Behaviour 85: 183-197.

ECHELLE, A.A. & ECHELLE, A.F. (1984) Evolutionary Genetics of a "Species Flock:" Atherinid Fishes on the Mesa Central of Mexico. In: Evolution of Fish Species Flocks, A.A. Echelle and I. Kornfield (Ed.). UMO Press, Maine, USA: 93-110.

FREYER, G. & ILES, T.D. (1972) The Cichlid Fishes of the Great Lakes of Africa. Their Biology and Evolution. TFH Publ. Neptune, N. J. USA

GEERTS, M. (1984) De Cichliden van de Westindische Eilanden. Period. Nederl. Cichl. Verenig.

GEERTS, M. (1988) Cichlidesque. Period. Nederl. Cichl. Verenig.

GREENWOOD, P.H. (1983) The Zoogeography of African Freshwater Fishes: Bioaccountancy or Biogeography? In: Evolution, Time and Space; The Emergence of the Biosphere. Sims, et al. (Eds.):179-201

GREENWOOD, P.H. (1984) African Cichlids and Evolutionary Theories. In: Evolution of Fish Species Flocks, A.A. Echelle and I. Kornfield (Ed.). UMO Press, Maine, USA: 141-154.

HASSE, J.H. (1981) Characters, Synonymy and Distribution of the Middle American Cichlid Fish *Cichlasoma meeki.* Copeia (1): 210-212.

HEIJNS, W. (1981) De Cichliden van het Nicaragua-Meer. Period. Nederl. Cichl. Verenig.

HEIJNS, W., GEERTS, M., 't HOOFT, J & KONINGS, A. (1985) Cichliden van de Wereld. Zuidboek Prod. Best, Holland.

MCKAYE, K.R. (1977) Competition for Breeding Sites between the Cichlid Fishes of Lake Jiloa, Nicaragua. Ecology, Vol. 58 (2): 291-302

MCKAYE, K.R. & BARLOW, G.W. (1976) Competition between Color Morphs of the Midas Cichlid, *Cichlasoma citrinellum,* in Lake Jiloa, Nicaragua. In: Investigations of the Ichthyofauna of Nicaraguan Lakes, T.B. Thorson (Ed.), Lincoln, Nebraska: 465-475.

KORNFIELD, I.L. & KOEHN, R.K. (1975) Genetic Variation and Speciation in New World Cichlids. Evolution, Vol. 29 (3): 427-437.

KORNFIELD, I.L., SMITH, D.C. & GAGNON, P.S. (1982) The Cichlid Fish of Cuatro Cienegas, Mexico: Direct Evidence of Conspecificity among Distinct Trophic Morphs. Evolution 36 (4): 653-664.

LIEM, K.F. & KAUFMAN, L.S. (1984) Intraspecific Macroevolution: Functional Biology of the Polymorphic Cichlid Species *Cichlasoma minckleyi.* In: Evolution of Fish Species Flocks, A.A. Echelle and I. Kornfield (Ed.). UMO Press, Maine, USA.

MILLER, R.R. (1966) Geographical Distribution of Central American Freshwater Fishes. Copeia, No.4: 773-802.

MILLER, R.R. & NELSON, B.C. (1961) Variation, Life Colors and Ecology of *Cichlasoma callolepis*, a Cichlid Fish from Southern Mexico, with a Discussion of the Thorichthys Species Group. Occ. Pap. Mus. Zool. Univ. Michigan, No. 622: 1-9.

MYERS, G.S. (1966) Derivation of the Freshwater Fish Fauna of Central America. Copeia, No.4: 766-773.

RIVERO, W. (1981) Taxonomic Status of Cichlid Fishes of the Central American Genus *Neetroplus.* Copeia (2): 286-296

ROSEN, D.E. (1976) A Vicariance Model of Caribbean Biogeography. Syst. Zool. 24: 431-464.

ROSEN, D.E. (1985) Geological Hierarchies and Biogeographic Congruence in the Caribbean. Ann. Missouri Bot. Gard. Vol. 72: 636-659.

SAGE, R.D. & SELANDER, R.K. (1975) Trophic Radiation through Polymorphism in Cichlid Fishes. Proc. Nat. Acad. Sci. USA, Vol. 72 (11): 4669-4673

STAWIKOWSKI, R. & WERNER, U. 1985. Die Buntbarsche der neuen Welt. Mittelamerika. Reimar Hobbing Verlag, Essen, FRG

UFERMANN, A., ALLGAYER, R. & GEERTS, M. (1987) Cichlid-Catalogue. Ufermann, Oberhausen, FRG.